POLYMER CONVERSION

POLYMER CONVERSION

W. A. HOLMES-WALKER

Director, The British Plastics Federation
London, England

A HALSTED PRESS BOOK

JOHN WILEY & SONS

NEW YORK — TORONTO

PUBLISHED IN THE U.S.A. AND CANADA BY
HALSTED PRESS
A DIVISION OF JOHN WILEY & SONS, INC., NEW YORK

Library of Congress Cataloging in Publication Data

Holmes-Walker, W. A. 1926–
  Polymer conversion.

  "A Halsted Press book."
  Includes index.
  1. Polymers and polymerization. 2. Plastics.
I. Title.
TP1087.H64   668.4'2   74–11510
ISBN 0–470–40767–0

WITH 49 TABLES AND 185 ILLUSTRATIONS

© APPLIED SCIENCE PUBLISHERS LTD 1975

Printed in Great Britain by Galliard (Printers) Ltd Great Yarmouth

PREFACE

This book is concerned primarily with the conversion of polymeric materials into useful articles. The subject is an immense one, and parts of it have been treated separately and exhaustively by many specialist writers. My aim has been to provide a rather simplified view of the whole sequence of operations starting with the raw materials, dealing in turn with the various processing stages and ending with the finished product.

We shall not concern ourselves with the pilot or industrial scale manufacture of polymers, since the techniques used are often highly complex and specific to individual materials. We are, however, concerned with the nature of polymers and their plan in the spectrum of materials, since by appreciating these things we are able to handle them in the most appropriate ways and for the most suitable applications. These points are also brought out again in the chapters on composite materials, and designing with plastics, which have been included in the text.

Also, since in most of the major conversion processes the polymer is subjected to various combinations of heat and pressure, a chapter on the rheological behaviour of polymeric materials is included.

After this we deal with the techniques used for the preparation of the polymer for the processing stages.

Processing itself is split up into what I have called the primary conversion processes, which involve the direct conversion of the powder or granules into shaped—but not necessarily finished—articles. This is followed by a consideration of the secondary conversion processes where a product—for example, a sheet—is further converted.

Obviously, this is only one of the many ways in which the matter could be treated, but it does have the merit of making it possible to emphasise two important considerations which run throughout the whole of the subject. They are the following:

(a) Although we have discussed different processes and parts of processes separately, the various stages of the conversion process are strongly interdependent and what happens at each stage will have a profound effect not only on subsequent stages, but also on the finished product.

(b) The processes themselves cannot either be considered in isolation since the choice of process will be influenced by such factors as:

(i) the material used;
(ii) the type of product;
(iii) the application;
(iv) the number of articles required.

It is for these reasons that chapters have been included on designing with plastics, plastics in combination with other materials, and sections on the layout of conversion plant.

I have provided only a very brief guide for further reading, since, as mentioned earlier, detailed information is best obtained by consulting the many specialist publications.

A large number of firms and individuals have been kind enough to provide illustrations and data, and where appropriate these have been acknowledged in the text. In addition it is a pleasure to place on record the help and information given by many friends and former colleagues.

Finally, I should like to express my gratitude to the publishers for their understanding and forbearance when this book had to compete, at times unsuccessfully, with many other tasks. A number of my colleagues have made helpful suggestions, and I am particularly indebted to Miss Alma West who has not only typed the work but has also provided much editorial help.

W. A. HOLMES-WALKER

CONTENTS

CHAPTER 1

POLYMERS AS MATERIALS

1.1 INTRODUCTION

Although this book is concerned primarily with polymers, it must at once be made clear that polymers are only a part of the whole spectrum of materials. It is obvious, therefore, that in order to be able to derive the maximum benefit from using polymeric materials it is absolutely vital that their place in the spectrum is thoroughly understood. This is particularly important since not only do the different classes of materials behave in strikingly similar ways, although usually under different time/temperature/pressure ranges, but processes originally designed for one material are being used or adapted for others. An example of this is the use of metal forging techniques in the solid phase forming of polymers.

Once this is done it is logical for us to take a closer look at polymers themselves and try to establish some of the factors which control their properties.

1.2 STRUCTURE/PROPERTY RELATIONSHIPS IN MATERIALS

All materials are made up from assemblies of atoms, and there are two principal factors which control the properties of the material. These are:

(a) the nature and variety of atoms;
(b) the arrangement of the atoms in space.

The importance of the nature of the atom in determining material properties is strikingly demonstrated by considering the two elements, carbon and nitrogen. These atoms are next-door neighbours in the periodic classification of the elements and their properties are shown in Table 1.1. We shall return to this theme later, but in the meantime we must consider briefly some other facets of atomic structure. This is such a complex subject with so many variations that there are dangers in attempting to

1

over-simplify. However, it is possible to observe a reasonably satisfactory connecting thread extending from relatively simple materials to those of considerable complexity.

TABLE 1.1

Element	Atomic number	Density at 20°C (g cm⁻³)	Melting point (°C)	Boiling point (°C)
C Graphite	6	2·25	3700[a]	4830
C Diamond	6	3·2	—	—
N Nitrogen	7	0·001 25	−210	−196

[a] Sublimes.

1.3 METALS

It is convenient to start with metals, not only because they are the most familiar engineering materials, but also since an appreciation of the fundamentals of their structure/property relationships will make it easier for us to understand polymeric behaviour.

Most metals consist of relatively simple, close-packed assemblies of atoms and it is possible for specific properties to be created in a number of ways.

1.3.1 The presence of imperfections in the crystal lattice

Engineering metals and alloys owe their unique combination of strength and ductility to their ability to slip. Practically all crystalline solids can slip, however—even diamond and sapphire when the temperature is high enough or when there is a sufficiently large hydrostatic pressure to prevent fracture. Some geological processes, for example, are brought about by the plastic deformation of normally brittle minerals under high pressures at great depths below the Earth's surface.

Where a fault or dislocation exists in the crystal there is a net increase in energy which renders it unstable. Under an applied stress this dislocation is able to travel through the material producing slip and thus deformation occurs relatively easily. The reduction in strength compared with that of the perfect crystal is often very striking in pure metals, and Fig. 1.1 shows how the strength disappears as soon as dislocations are made in a copper whisker.

The calculated strength of metals should be about $E/20$ whereas it can be seen from Fig. 1.1 that once dislocations have been created the stress has fallen to about $E/2000$. The initial yield stress, incidentally, in the case

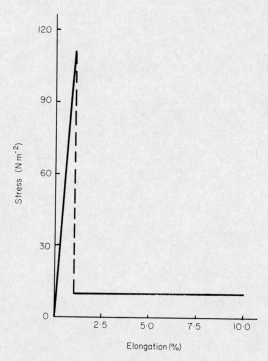

FIG. 1.1. Stress–strain curve of a copper whisker.

of a dislocation-free copper whisker ($E/200$) is of the same order as the tensile strength which can be obtained for copper by extreme cold-working at room temperature.

The creation of imperfections by the addition of foreign atoms generally serves to strengthen the material. The solute atom decreases the stress around a dislocation; as a result the dislocation is more stable and requires a greater stress to move it. Thus solid solution metals always have higher strengths than do pure metals (Fig. 1.2).

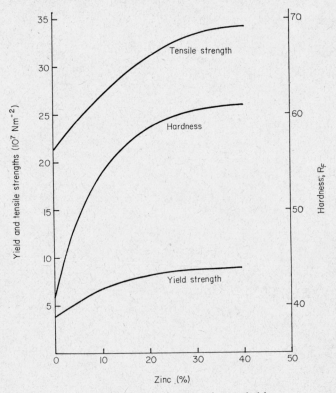

FIG. 1.2. Mechanical properties of annealed brasses.

1.3.2 The effect of grain structure

A more massive re-arrangement of the crystal structure of single phase metals can produce a structure consisting of a variety of crystals of different orientations. The adjacent crystals have dissimilar orientations so that a grain boundary is present (Fig. 1.3). The microstructures of single phase metals can be varied by changes in size, shape and orientation of the grains, and the grain structure will influence the mechanical properties of the metal. For example, the changes in ductility and tensile strength in an annealed 70/30 brass (Fig. 1.4) are a direct reflection of the grain boundary area within the brass and the effect that the boundary has on slip.

1.3.3 The introduction of a second phase

This can often have a strengthening effect by inhibiting the movement of dislocations. To illustrate this it is useful to consider the Fe–C phase

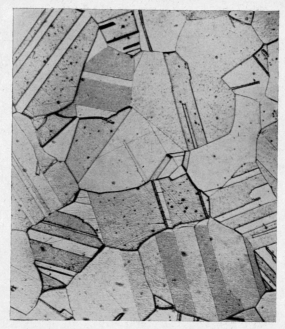

FIG. 1.3. Illustration of grain structure in a recrystallised 70/30 brass (×100).

relationships since steel offers examples of the majority of reactions and microstructures available for adjusting material properties. We are also able to bring in another parameter—temperature—which, as we shall see later, has a profound effect on all materials.

TABLE 1.2

Effect on steel properties of varying the character of the disperse phase

Composite system			Disperse phase variables changing:		
Phase					*Concen-tration*
			Shape	*Size*	
Continuous	*Disperse*	*Property*	*Plate to round to fibre*	*Increasing*	*Increasing*
Iron	Carbides	Toughness	Increase	Decrease	Increase
		Hardness	Decrease	Decrease	Increase

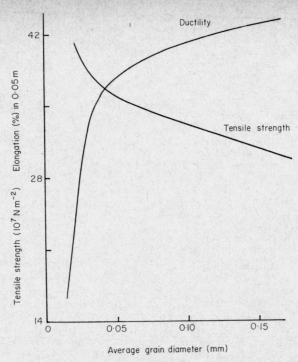

Fig. 1.4. The dependence on grain size of the mechanical properties of an annealed 70/30 brass.

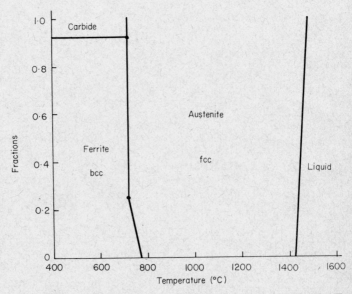

Fig. 1.5. Phase fractions *vs* temperature for an alloy of 0·6 C: 99·4 Fe.

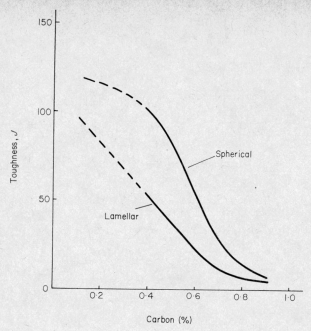

FIG. 1.6. Effect of carbon concentration and particle shape on the toughness of steel.

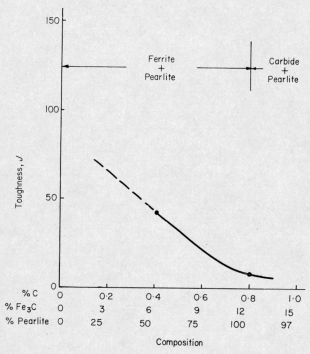

FIG. 1.7. Toughness *vs* composition of steels.

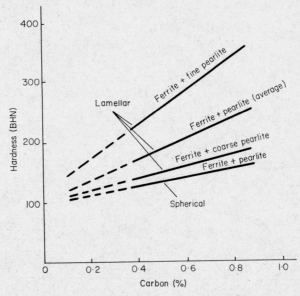

Fig. 1.8. Effect of particle size and shape and percentage carbon on hardness of steel.

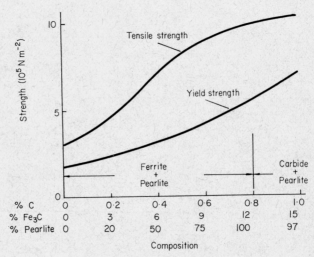

Fig. 1.9. Strength *vs* composition of steels.

Pure iron changes its crystal structure from body-centred to face-centred cubic as it is heated above 910°C and a subsequent phase change occurs at 1400°C. These changes and their effects on the solubility of carbon in iron are shown in Fig. 1.5.

In iron–carbide alloys, carbon in excess of the solubility limit must form a second phase, and this second phase has the chemical composition Fe_3C. The most familiar of the resulting structures is known as Pearlite; the mixture is lamellar and composed of alternate layers of ferrite and carbide. Pearlite is a specific mixture of two phases formed by transforming austenite of eutectoid composition to ferrite and carbide. However, it is possible to form mixtures by other reactions, although these microstructures will not be lamellar and consequently their properties will be different. Two other parameters will also affect the properties of these composite materials; these are the volume fraction and the size of the dispersed phase. A brief survey of the effects of these parameters is shown in Table 1.2 and Figs. 1.6, 1.7, 1.8 and 1.9.

Having considered metals and their properties in some detail we can now appreciate the wide variety of ways in which it is possible not only to select appropriate atoms but also to control their arrangements in space in order to create a whole range of physical properties.

1.4 CERAMICS

Ceramic materials contain phases which are compounds of metallic and non-metallic elements. There are many ceramic phases because:

(a) there are many possible combinations of metallic and non-metallic atoms;
(b) there may be several structural arrangements of each combination.

Most ceramic materials, like metals, have crystal structures. The electrons, however, are either shared covalently with adjacent atoms, or are transferred to produce ionic bonds. For this reason ceramic materials are in general highly stable. As a class they have higher melting temperatures than do metals or polymeric materials and they are also harder and more resistant to chemical attack. In some of the simpler crystals such as MgO, plastic slip, similar to metallic slip, can occur. Crystal outlines can form during growth as exemplified by the cubic outline of common salt.

In asbestos the crystals have a marked tendency towards linearity; in micas and clays the crystals form two-dimensional sheet structures and the stronger, more stable ceramic materials commonly possess three-dimensional framework structures with equally strong bonding in all three directions.

1.5 POLYMERS

We have seen that, in the case of metals and ceramics, the atom, the ion, and the unit cell are the important features of structure in determining properties. In the case of polymers, however, the polymer chain is the important feature, and it is the nature of the chain links (their size and shape), together with the ways in which the links are arranged to form the chain, the length of the chains, and the disposition of the chains in space, that determine the nature of the material. The simplest type of polymer—which we may describe as a one-dimensional macromolecule—consists of a single chain of backbone atoms and the most familiar member of this class is polyethylene, which is composed of ethylene molecules.

We can gain an idea of the size of the polyethylene chain if we appreciate that the length of the ethylene molecule is about 2×10^{-10} m long and that a typical polyethylene chain of some 50 000 links would be about 2×10^{-5} m long (in the region of a fiftieth of a millimetre).

The polyethylene chains are normally tangled and coiled in random directions like the filaments in a bundle of wire wool. By looking at Fig. 1.10(a) we see that the material is flexible because rotation can easily occur at each of the bonds, hence the molecule is readily deformed. It is tough because this movement will absorb sudden stresses. It is light because there is a large space/chain ratio (this is often increased by considerable branching of the main chain structure) and it has a low softening point because the chains can easily disentangle and slide past one another when thermal energy is supplied to the material. This latter characteristic, as we shall see later, is common to all thermoplastic materials which, providing excessive heat is not used, can be reprocessed.

1.5.1 Methods of creating specific properties

The flexibility of the polymer molecule, and hence the material, can be controlled by a variety of methods. It can be increased by the addition of plasticisers. Plasticisers, in simple terms, are liquids which fill up some of the spaces between the chains and so make it easier for them to move about relative to one another. This technique is most frequently used in PVC technology. Flexibility can also be increased by the technique of copolymerisation, whereby some of the polymer links in the chain are replaced by ones which can more easily be deformed.

(*a*) *Substitution in the chain*

Flexibility can be decreased by introducing a series of knobs or lumps into the chains thus restricting their movement by making it difficult for the links to rotate. The substitution of the bulky chlorine atoms and the even

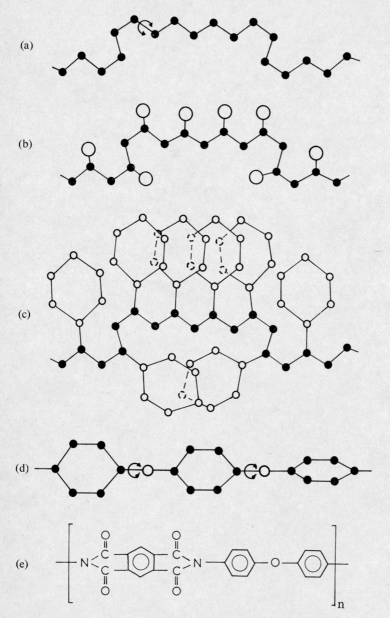

FIG. 1.10. Polymer molecules of increasing chain stiffness. (a) Polyethylene. (b) Rigid PVC. (c) Polystyrene. (d) Polyphenylene oxide. (e) Polyimide.

Polymer Conversion

larger benzene rings for some of the hydrogens in the chains (see Figs. 1.10(b) and 1.10(c)) produces, respectively, rigid PVC and polystyrene. Both these materials are stiffer than polyethylene but are also brittle since the lumps prevent the dissipation of stresses by chain movements. Also, as the chains are separated more widely by these bulky side groups, both materials have relatively low softening points and are easily attacked by solvents. By allowing the chains to rotate but not bend, it is possible to confer good mechanical and thermal properties on the material. Examples are polyphenylene oxide and polyimide (Figs. 1.10(d) and 1.10(e)).

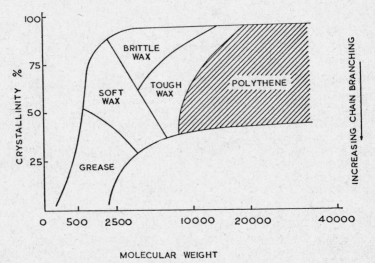

Fig. 1.11. Effect of molecular weight on materials properties.

(b) Chain arrangements in space

Stronger, stiffer materials can also be created by a number of other techniques. Increasing the lengths of the polyethylene chains from a few hundreds to many thousands of links produces the variety of materials shown in Fig. 1.11. By eliminating the branches in the chain and making them lie closer together, we can produce a partly ordered or crystalline structure in which the density has increased from about $0 \cdot 92$ g cm^{-3} to more than $0 \cdot 96$ g cm^{-3}. One of the characteristics of this type of structure is that the same molecule may exist in an ordered crystalline region at one point, and in an amorphous region at another. Arguments continue about the fine structure of the crystalline and amorphous regions, but there is broad agreement on the general picture of the molecular arrangement, which is illustrated in Fig. 1.12.

As well as alterations in the chain length (molecular weight), the control of molecular weight distribution, in which the spread of chain lengths is concerned, also provides an important means of determining properties.

The amount of ordering of the polymer chains can also be increased by orientation. The effect of carrying out this process of orientation on polystyrene is shown in Table 1.3. Although monoaxial orientation is carried out in order to increase the strengths of textile fibres, biaxial orientation of films also provides a considerable increase in strength by orienting the molecules in the plane of the film.

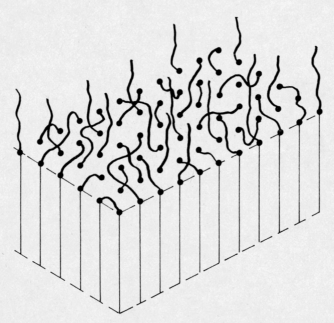

FIG. 1.12. Illustration of the morphology of a crystalline polymer crystallised from bulk solution. In the lamellar structures shown, the vertical lines represent ordered chains which are either in planar or zigzag configurations or in one of the helical forms. The random lines represent the amorphous regions between crystallites.

In the film-making process the polymer at various stages in its passage through the extruder and die will have been heated above its melting point, and so much of its crystalline structure will have been destroyed. The molecules will have been made to disentangle and slide past one another to take up new positions. On cooling, the molecules can once again settle down into a partially ordered arrangement as before. Provided, in the case of tubular film manufacture (cf. Chapter 4), the distension of the bubble

TABLE 1.3

Strengths of orientated films and fibres

Material	Draw ratio	Stress at break $(10^7 \ N \ m^{-2})$	Type of orientation
Polystyrene	1:1	3·5	Uniaxial
(film)	3·75:1	8·3	Uniaxial
	10:1	69	Uniaxial
Polyethylene	1:1	0·7	Uniaxial
(film)	14:1	138	Uniaxial
Polyethylene	1:1	4·1	Biaxial
terephthalate	2:1	8·2	Biaxial
(a) 'Melinex' film	3·5:1	21	Biaxial
(b) Fibre	6:1	40	Uniaxial
	7:1	55	Uniaxial

occurs during this 'molten' state, the mechanical properties of the film will be little altered from those of the starting polymer.

However, if we allow the material to cool before stretching to the extent that the crystalline regions have at least partially reformed, then the only parts of the material able to yield are the tangled amorphous regions. Figure 1.13 shows this process in a very much simplified form. The crystallites are represented by the heavy lines (Figs. 1.13(a) and 1.13(b)) and it can be seen that in the unorientated material the crystallite can be in almost any direction. When the film is stretched forwards and sideways (x and y axes) the crystallites pivot round the amorphous links. In order to compensate for the increases in x and y, z becomes smaller, i.e. the material becomes thinner (Figs. 1.13(c) and 1.13(d)). Although when viewed from above the orientation of the crystallites seems little changed, the z contraction has caused them to tilt and arrange themselves more in the plane of the film (Fig. 1.13(d)). Since the amorphous regions take part in this process, the tangled chains tend also to become more ordered, and therefore there will be an increase in density as well as an increase in strength.

If, after orientation, the temperature of the film is raised above the stretching temperature, the restoring forces will cause the film to retract almost to its original dimensions. This is the process which operates in the 'shrink-wrap' films. On the other hand, thermal stability can be achieved in, for example, 'Melinex' polyester film by a further heat-setting or crystallisation stage in which the film is maintained under tension after the biaxial orientation. The increase in mechanical properties resulting from this orientation and heat-setting on 'Melinex' polyester is shown in

Table 1.3. Without heat-setting there will be a tendency for the orientation to be lost as in the shrink-wrap films. The amount of reversion in relatively highly crystallisable materials like 'Melinex' is smaller due to the strain-induced crystallinity caused by stretching at the higher ratios. We can say, however, that orientation and subsequent heat-setting at 220°C increases the temperature at which the material exhibits 5% shrinkage from about 80°C (the glass transition temperature) to about 200°C. There are other

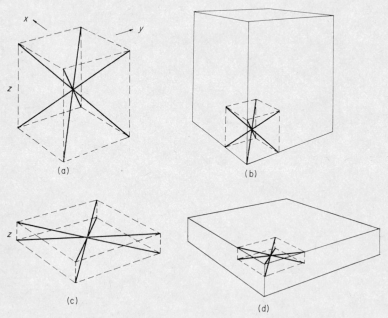

FIG. 1.13. A schematic representation of orientation in polymers. (a), (b) Part of an unoriented sample of material. (c), (d) Uniformly oriented material showing the increase in size accompanied by a decrease in thickness.

problems associated with high temperature where sustained use at 200°C is likely to lead to chemical degradation unless measures are taken to prevent it.

(c) The introduction of struts or cross-links

We can go even further in pursuit of strength and rigidity by incorporating a series of struts in the polymer molecule, thus producing a chain that will neither bend nor rotate. This technique is used in the production of the so-called 'ladder polymers', an example of which is shown in Fig. 1.14.

This technique is used in the production of one type of carbon fibre—the pyrolysis of polyacrylonitrile—and gives a material which can be brought to red heat without damage. At these temperatures there will be a tendency for materials to lose strength by oxidation or by hydrolysis upon prolonged exposure, but provided these limitations are appreciated the methods I have described do provide a range of most valuable engineering materials.

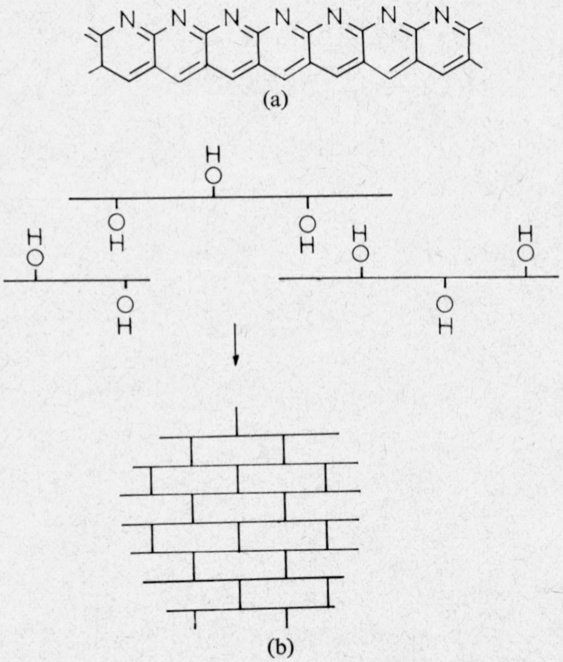

(a)

(b)

FIG. 1.14. Ladder and cross-linked polymers. (a) Ladder polymer: 'black Orlon'. (b) Formation of an alkyd resin.

Finally, we can join up all the chains by other links and so obtain a rigid three-dimensional network, shown diagrammatically in Figs. 1.14(b) and 1.15. By appropriate choice of the molecular links in the chains, and by controlling the nature and number of the cross-links, we are able to design materials with a very wide range of properties. The vulcanisation of rubber—shown diagrammatically in Fig. 1.16—gives an illustration of the effect of altering the cross-link density. Young's modulus values range from $2 \times 10^6 \, \text{N m}^{-2}$ for unfilled lightly vulcanised material, to about $3 \times 10^9 \, \text{N m}^{-2}$ for ebonite containing about 40 parts of sulphur per 100 parts of rubber.

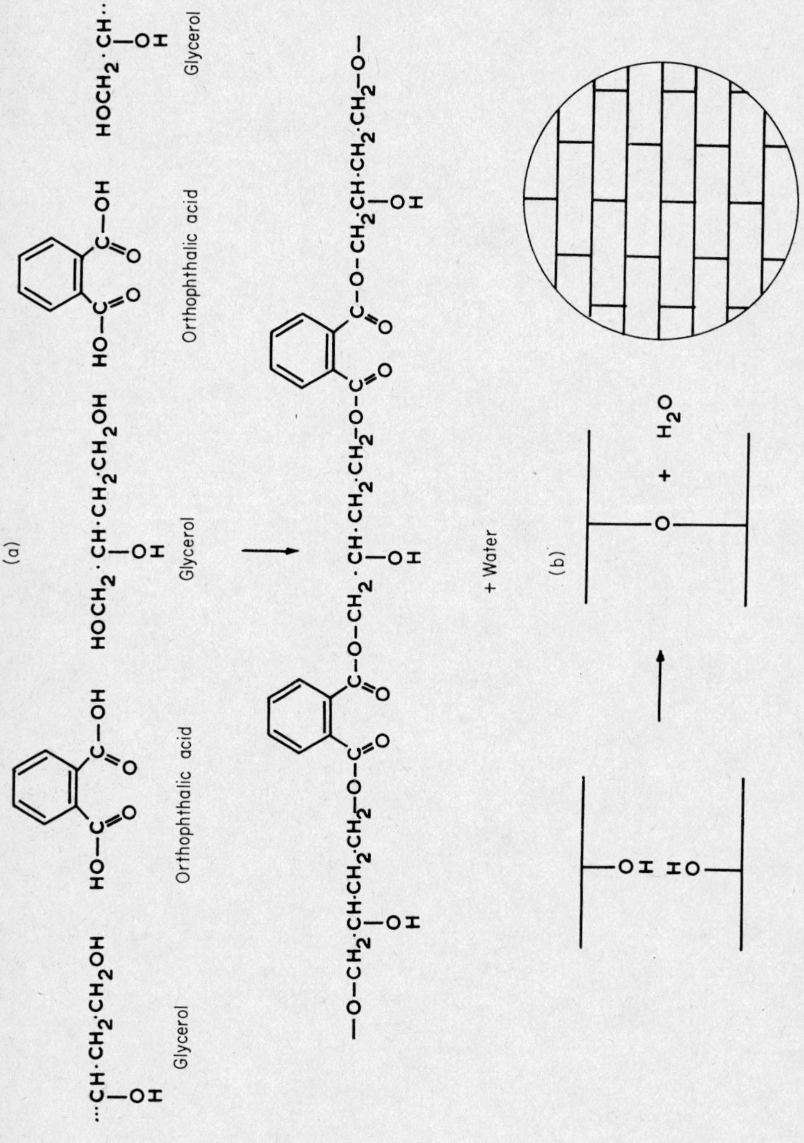

FIG. 1.15. Formation of an alkyd resin. (b) Subsequent network formation.

Polyesters, on the other hand, provide an example of the variety which can be achieved by the use of different molecular links (Figs. 1.15 and 1.17). By choosing the size, shape and chemical nature of the three components—

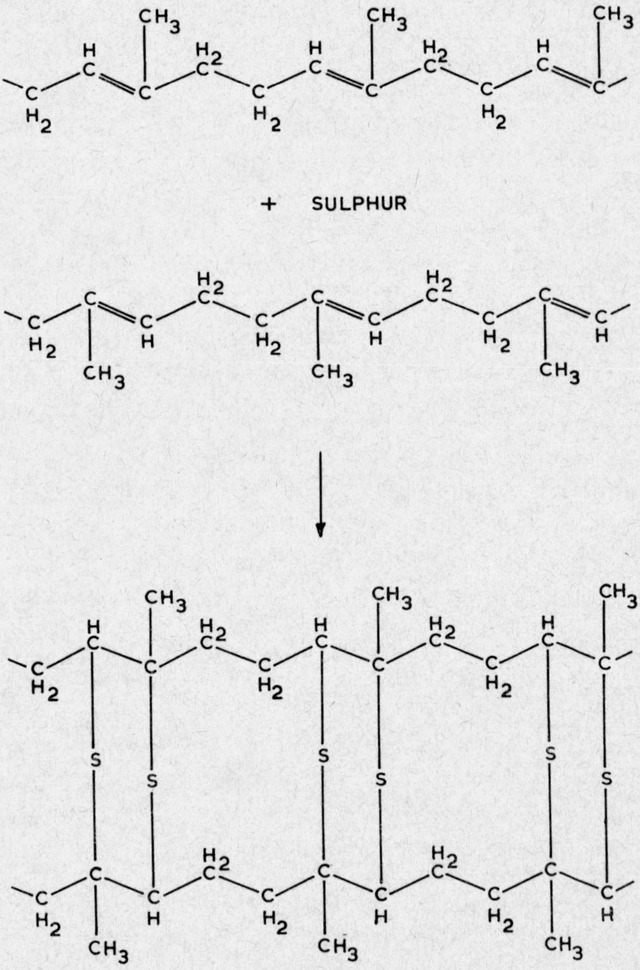

+   SULPHUR

FIG. 1.16. Cross-linking of rubber.

acid, alcohol and cross-linking monomers—it is possible to 'design-in' almost any desirable property. For example, in the case of the alcohol, branched or bulky groups produce resins that are tougher and have higher softening points than those made with symmetrical alcohols. Oxygen linkages increase the water sensitivity until, with heptaethylene glycol for instance, we obtain a material which is partly water-soluble. The long flexible chain also increases the flexibility of the cured resin. On the other hand, polyesters having outstanding chemical resistance and considerable mechanical strength can be achieved by the use of a bulky but stiff molecule of the type shown in Fig. 1.10(d).

Similar variations in structure can, of course, be incorporated in the acid component, although some of the examples are more expensive than

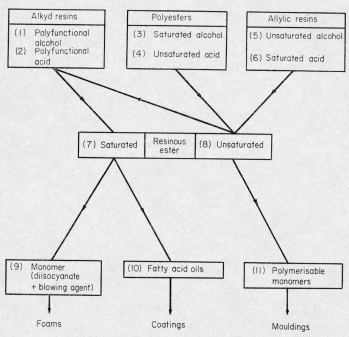

FIG. 1.17. Illustration of the relationships between different types of 'polyesters'.

is the case with the alcohol. The use of chlorinated acids, however, has proved useful in reducing the flammability of the polyester resins.

Styrene has for many years been the traditional cross-linking monomer as it is cheap, and resins based on styrenes cure rapidly and have good weathering characteristics. Unfortunately, the refractive index of styrene-modified polyesters is high (1.54). Since this is higher than that of glass (1.51), roof sheeting based on styrene-modified resins shows a strong glass-fibre pattern. By a fortunate piece of scientific juggling it was found that not only are the glass fibres almost invisible in polyesters containing mixed methacrylate and styrene monomers, but these materials also had improved weathering resistance. In addition the presence of the methacrylate retards the cure of the polyester, a feature of some value where continuously operating processing sequences are concerned. α-Methyl styrene has been used as the monomer in order to reduce the shrinkage on curing as well as

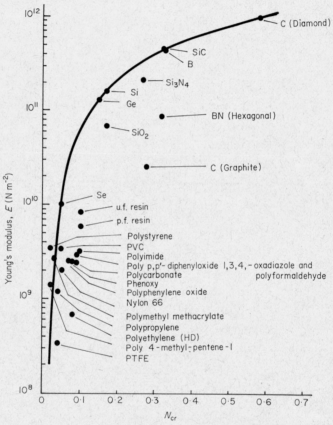

FIG. 1.18. Young's modulus (E) plotted against N_{cr} for inorganic and organic polymer systems.

to provide resins with enhanced flexibility. Polyesters having considerable heat resistance can be obtained by the use of triallyl cyanurate as the modifying monomer.

The adoption of rigid three-dimensional networks is, incidentally, a trick used by nature to create such immensely strong materials as diamonds and silica. This increase in strength as we go from linear to condensed systems is well illustrated by Fig. 1.18 which shows the variation of Young's modulus with N_{cr} (a structural parameter which is defined as the relative number of network bonds per unit volume) for a variety of covalently bonded polymers.

To sum up briefly, we have considered the following techniques of structural manipulation: (a) chain modification; (b) chain lengthening; (c) chain strengthening; (d) chain stiffening; (e) orientation; (f) cross-linking; and seen that they are all able to promote desirable properties in the material. Table 1.4 shows, for example, the effect of carrying out some of these modifications on the maximum service temperatures of a range of plastics.

TABLE 1.4

Maximum service temperatures of plastics

Material	Chain structure	MST ($^{\circ}C$)
Polyethylene	Flexible	120
Polycarbonate	Rigid	135
Nylon 66	Rigid	150
Phenolic	Cross-linked	175
Epoxy	Cross-linked	235
Polyimide	Very rigid	300

Most of the techniques involving the molecular chains which we have discussed so far are used to 'improve' polymer properties by increasing their resistance to the environment and conferring greater mechanical strength. These admittedly are the usual requirements for 'engineering materials'. However, there are applications where such properties as transparency, flexibility and increased toughness, as distinct from rigidity, are required.

(d) The importance of isomers

There exists another most important structural technique, which I have left until this stage. This is the use of stereoisomerism. Originally discovered by Pasteur in the field of small organic molecules, when applied to polymers it is capable of producing materials with widely differing properties.

Stereoisomers are molecules possessing identical structural formulae, but which have a difference in the distribution of atoms and groups in space. These isomers can exist because a restriction, such as for example a double bond, prevents the tendency of the atoms to revert to a common arrangement.

One of the simplest examples of stereoisomerism is shown below:

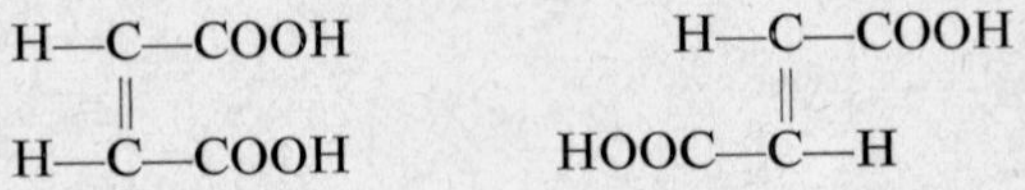

$$H—C—COOH \qquad\qquad H—C—COOH$$
$$\parallel \qquad\qquad\qquad\qquad \parallel$$
$$H—C—COOH \qquad\qquad HOOC—C—H$$

maleic acid (*cis*)   fumaric acid (*trans*)

These two compounds are stereoisomers, and the descriptions *cis* and *trans* refer to the disposition of the chemical groups—on the same side and across respectively—about the double bond.

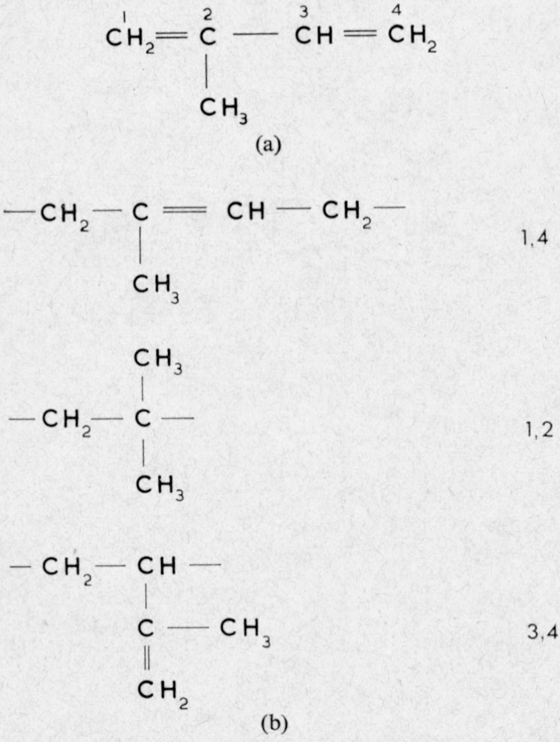

$$\overset{1}{C}H_2 = \overset{2}{C} — \overset{3}{C}H = \overset{4}{C}H_2$$
$$\mid$$
$$CH_3$$

(a)

$$—CH_2— C = CH — CH_2— \qquad 1,4$$
$$\mid$$
$$CH_3$$

$$CH_3$$
$$\mid$$
$$—CH_2— C — \qquad\qquad\qquad 1,2$$
$$\mid$$
$$CH_3$$

$$— CH_2— CH — \qquad\qquad$$
$$\mid$$
$$C — CH_3 \qquad\qquad 3,4$$
$$\parallel$$
$$CH_2$$

(b)

FIG. 1.19. The polymerisation of isoprene. (a) The isoprene molecule. (b) Type of addition.

Many polymer molecules also contain double bonds, and in particular polyisoprene provides a useful illustration of a number of additional ways in which the isoprene units can form stereoisomers. The isoprene molecule is shown in Fig. 1.19(a). The most familiar method of chain formation is by 1,4 addition and this, together with the other types of addition, is also shown in Fig. 1.19.

Stereoisomers are produced by using a special type of catalyst. This catalyst, by means of its highly specific arrangement of positive and negative charges, acts as a jig or template and ensures that the reacting molecules can only link up in a certain way.

(e) Copolymerisation

We have already seen that flexible materials can be created by using plasticisers. Another most important method of designing properties into the molecular chains is that of copolymerisation. Here, different chain units, let us call them A and B, are: (i) distributed randomly throughout the chains; (ii) in blocks of A and B; or (iii) as grafts of A onto a backbone composed entirely of units of B.

With two monomers (A and B) it is possible to obtain a range of polymers with characteristics that vary broadly according to the proportions of each. With more than two monomers, the possibilities are, in theory at least, considerably larger. However, control of the variables of the polymerisation process presents some problems and, for instance, polymers of acrylonitrile butadiene styrene are disappointing in that toughness is only achieved at the expense of other properties. Greater success is achieved by the incorporation of a lightly cross-linked elastomeric phase (e.g. butadiene) into a copolymer of styrene/acrylonitrile (see also Chapters 3 and 8).

1.6 POLYMERS COMPARED WITH OTHER MATERIALS

Having built up a picture of how we can create properties in polymers we can now compare them with other classes of materials with which they might compete. We must remember, however, that some of the techniques we have considered lead to the production of highly sophisticated materials with rather narrow fields of application. Therefore, to simplify the comparison we shall at this stage consider only relatively simple polymers.

Figure 1.20 shows the tensile strengths of a range of materials with metals occupying about a decade centred on 10^9 N m^{-2}. The spread for normal plastics extends from above 10^8 down to 10^7 N m^{-2} and the various foamed polymeric materials occupy the region below this down to about 10^5 N m^{-2}. This last need not worry us unduly since concrete is an excellent engineering material of poor tensile strength.

With rigidity, we find (Fig. 1.21) that most of the familiar materials of construction are grouped slightly above a Young's modulus of 10^{11} N m^{-2}. Diamond and carbon whiskers are slightly below 10^{12} N m^{-2}, and the span of plastics goes from 10^{10} to 10^7 N m^{-2}. Although, as we shall see later, it is possible to alter many of the physical properties of the material by selection of the appropriate processing techniques, it is difficult to affect

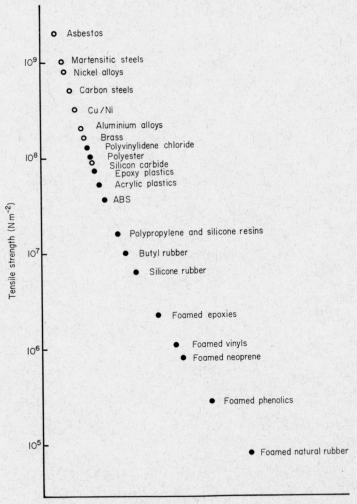

FIG. 1.20. Tensile strengths of a range of materials.

rigidity—except by using materials in combination—since this characteristic is determined by the way in which the component parts of the molecule are put together.

Figure 1.22, which shows the melting points of materials, emphasises the point made earlier that the different physical responses of materials occur over a wide range of temperatures. Many plastics, and in particular

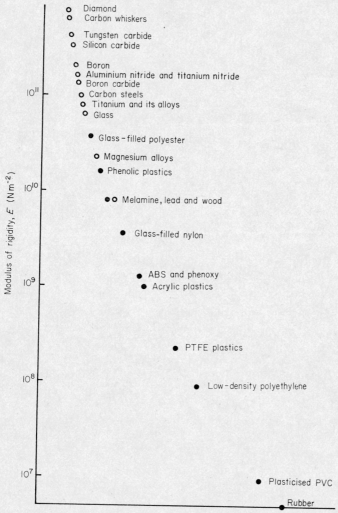

FIG. 1.21. Modulus of rigidity of a range of materials.

the reinforced materials, though both strong and rigid have poor thermal resistance; only the most sophisticated are usable above 500°C and the average is well below 300°C.

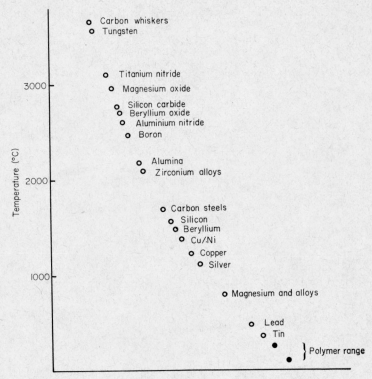

FIG. 1.22. Melting points of various materials.

1.7 USING PLASTICS

The picture presented by these data may not at first sight appear too encouraging. However, several factors must be borne in mind:

(a) they are relatively light, and if we compare specific strengths—obtained by dividing the strengths by their specific gravity—the picture alters considerably;
(b) they are often extremely flexible as well as tough;
(c) many plastics are highly transparent;
(d) many identical, and often complex, articles can be made from one mould;
(e) the number of components and assembly stages can often be reduced by good design;
(f) little 'finishing' is required;
(g) colour can be 'built in'.

We shall see in a later chapter that the simple comparisons given in Figs. 1.20, 1.21 and 1.22, although useful as a guide, are rather naive. Before we can develop this idea further though we must turn to the behaviour of polymers under applied stresses.

FURTHER READING

Anderson, J. C., and Lever, K. E. (1969). *Materials Science*, Nelson, London.
Cottrell, A. H. (1964). *The Mechanical Properties of Matter*, Wiley, London.
Kaufman, M. (1968). *Giant Molecules*, Aldus Books, London.
van Vlack, L. H. (1970). *Materials Science for Engineers*, Addison-Wesley, London.

CHAPTER 2

FLOW BEHAVIOUR IN POLYMERS

2.1 INTRODUCTION

During their conversion into useful articles polymers undergo a sequence of processing operations all of which will, to a greater or lesser degree, affect the properties of the finished product. The principal variables in this sequence are time, temperature, pressure and also, to some extent, the environment.

The ways in which these parameters operate depend not only upon the process being used, but also on the nature of the polymer, and the most important, in terms of final product properties, are the physical effects of mixing and orientation and the chemical effects of degradation.

We saw in the previous chapter how the properties of polymers can be profoundly influenced by changes in structure, and it is now necessary to attempt to understand the effect of processing on polymer performance. This is important for the better exploitation of existing processes, and also to be able to design new processes capable of meeting new requirements and handling new classes of materials.

2.2 RESISTANCE TO DEFORMATION

One of the characteristics of polymers is that their rheological behaviour has a dual nature; at times they show the features of elastic solids and at others they behave like viscous liquids. Most polymeric materials at some stage in their responses display both these characteristics, and are described as 'viscoelastic'.

Although the term viscoelasticity is used when the whole range of behaviour is being considered, a sub-division is sometimes introduced where materials are termed viscoelastic when they are essentially elastic solids which show some slight viscous behaviour. This is particularly so when we

28

attempt to understand the response of polymers to dynamic stimuli; a type of response which is becoming increasingly important in assessing the 'engineering' characteristics of plastics and rubbers.

During processing, however, polymers essentially behave as fluids but may show some elastic effects—such as die-swell (cf. Section 2.8.1) in extrusion—and this behaviour is increasingly known as 'elastico-viscous' behaviour. Although in this chapter, as an introduction to processing, we are principally concerned with the 'elastico-viscous' components of polymer behaviour, we shall in the development of mechanical models deal with the whole spectrum of responses. This is necessary for two reasons; firstly, since the boundaries of behaviour patterns are not clear and often overlap, it is better to develop the picture as a whole—at least in outline—and then make use of the points which we need. Secondly, since not only the evaluation of polymers but also in some instances their conversion occurs in the 'viscoelastic' regions of response, it is necessary to be familiar with this part of the picture.

There is one common factor, however, which operates over the whole range of materials from solids to liquids; the application of a load or shear causes deformation or strain. This deformation may occur instantaneously or it may continue with time. Many models have been used to illustrate this type of behaviour, and we shall discuss several of them at a later stage. The most complex obviously approach most closely to reality, but all must exhibit the following basic behavioural patterns observed when a polymer is stressed.

On application of the stress there follows:

(a) an instantaneous elastic deformation (owing to the bending and stretching of primary valence bonds);
(b) a retarded and recoverable elastic deformation (owing to the molecular configuration moving to the new equilibrium position associated with the elongated or oriented state);
(c) an irrecoverable deformation (owing to the polymer chains or segments slipping past one another).

The relationships between stress, strain and time can be expressed mathematically and are known as the rheological equations of state.

It will be useful now to look at some of the equations which have been developed to explain the behaviour of materials. We must, however, remember that they usually refer to idealised materials and that the behaviour of real materials—especially under real conditions—is often very different. This applies particularly when we attempt to use with any precision the mathematical models which have been developed for such complex processes as, for example, extrusion film blowing and calendering. Provided the limitations are appreciated and understood, though, such models do provide some useful guidance.

2.3 MECHANICAL PROPERTIES

2.3.1 Elastic deformation

A material is called elastic when the deformation produced in the body is wholly recoverable after the removal of the force. The relation between stress and strain in the elastic region of the material is governed by Hooke's Law, which states that stress is proportional to strain, and independent of time. This law generally applies to most elastic materials for very small strains. For an isotropic material each stress will induce a corresponding strain, but for an anisotropic material a single stress component may produce more than one type of strain in the material. This phenomenon is particularly important in the case of the 'real' materials we encounter in use, since processing and fabrication methods almost invariably introduce anisotropies into materials.

2.3.2 The ideal solid

This material obeys Hooke's law and it follows that the ratio of stress to strain is a constant characteristic of the material. This proportionality constant is called the modulus of elasticity. Since there are three main types of stress: tension, compression and shear, there will be three corresponding moduli of elasticity:

(a) the modulus of elasticity in tension, or Young's modulus, usually denoted by E is represented by:

$$E = \sigma_t/\varepsilon_t \tag{2.1}$$

where σ_t is the tensile stress and ε_t the tensile strain;

(b) the modulus of compressibility or bulk modulus, denoted by K, is defined as the ratio of hydrostatic pressure to the relative change in volume. The resultant stress in the material is the effect of three stresses acting in the three principal directions at right angles.

$$K = \frac{\sigma}{\Delta V/V_0} = \frac{\text{stress}}{\text{volumetric strain}} \tag{2.2}$$

when σ is the stress, and $\Delta V = (V_0 - V)$;

(c) the modulus of rigidity or shear modulus, denoted by G, is

$$G = \tau/\gamma \quad \text{or} \quad \sigma/\gamma \tag{2.3}$$

where τ or σ is the shear stress and γ the shear strain. An important elastic constant is Poisson's ratio ν; it is defined as the ratio of the lateral contracting strain to the elongation strain when a body is stretched by a force applied at its ends.

Any two of these constants may be regarded as fundamental and the other two may be calculated from them. For isotropic systems and small deformations these four elastic constants are related to one another.

2.3.3 Relations between moduli

The various elastic constants are often determined by different experimental techniques, some of which may be difficult to use or even quite inappropriate for particular samples of material. It is therefore useful to know how these constants are related to one another in order to be able to extract the maximum amount of useful information.

(a) The bulk modulus K is related to the Young's modulus E as follows:

$$K = \frac{E}{3(1 - 2\nu)} \tag{2.4}$$

where ν equals Poisson's ratio.

(b) The modulus of rigidity or shear modulus G is related to the Young's modulus E as follows:

$$G = \frac{E}{2(1 + \nu)} \tag{2.5}$$

(c) Young's modulus E is also related to both the other constants G and K as follows:

$$E = \frac{3G}{1 + G/3K} \tag{2.6}$$

or

$$\frac{1}{E} = \frac{1}{GK} + \frac{1}{3G} \tag{2.7}$$

Since $3K$ is usually much larger than G we can write

$$E = 3G\left(1 + \frac{G}{3K}\right)^{-1} \simeq 3G\left(1 - \frac{G}{3K}\right) \tag{2.8}$$

This shows that Young's modulus is contributed to mainly by the shear modulus and only slightly by the bulk modulus.

The behaviour of a Hookean solid can be described graphically as shown in Fig. 2.1. In Fig. 2.1(a) a stress is applied at time t_1. A strain occurs instantaneously and remains constant until the stress is removed at time t_2. At this moment also the strain drops to zero.

Since the elastic recovery is due to the action of interatomic and intermolecular forces, the values of the elastic moduli will vary directly with the magnitude of these forces. Thus the moduli of elasticity of such covalent compounds as diamonds are very high—they are lower for metallic and ionic crystals. Molecular amorphous solids such as plastics and rubbers and molecular crystals show relatively low values for their moduli of rigidity (see Fig. 1.21).

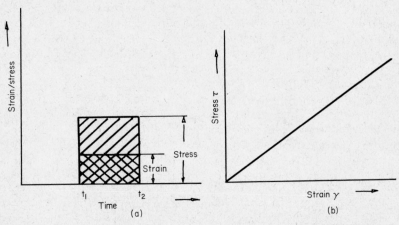

FIG. 2.1. Stress–strain–time relationships for an ideal solid.

2.3.4 The effect of temperature

As we might expect, there is also a decrease in the values of the elastic moduli as the temperature increases, since thermal energy helps to overcome the interatomic forces. With metals the change is slow at first, increasing rapidly as the melting point is approached.

The elastic moduli of plastics and other amorphous materials are even more temperature sensitive. An exception occurs with rubbers and other elastomers because of their kinked conformation, and we shall consider this in more detail later.

This drastic alteration which occurs in the behaviour of polymers under stress over the range of temperature from below zero to about 200°C is of enormous practical importance, and can tell us a great deal about the contribution of structure in the response of materials to mechanical stress.

Figure 2.2 shows a typical relationship between modulus and temperature. The case of polystyrene is shown in rather more detail in Fig. 2.3.

2.4 FLUID FLOW: IDEAL BEHAVIOUR

Following the relationships derived in the last section, we can make the following brief generalisations.

(a) For most soft materials, such as colloidal systems, gels, pastes and

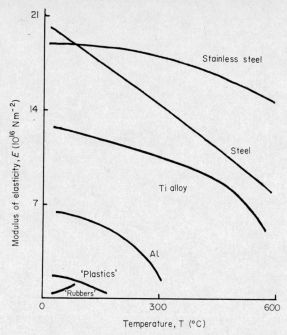

FIG. 2.2. Variation of elastic modulus with temperature.

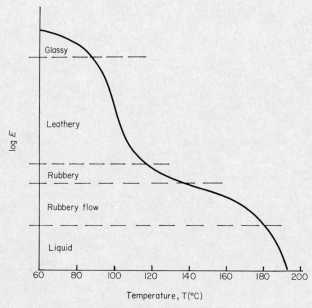

FIG. 2.3. Relationship between modulus and temperature for polystyrene.

putties, K is very large compared with G, and it follows from eqn. (2.7) that $E = 3G$. Such materials are regarded as incompressible because the stress applied usually causes a change in shape in preference to compression.

(b) In the case of liquids, which can sustain only a hydrostatic pressure for any length of time, the bulk modulus K is the only elastic modulus of importance.

Turning now to specific examples of material behaviour, it is convenient to discuss the different types of response to stress in turn.

2.4.1 The Pascallian fluid

Such a material obeys Pascall's law in that pressure applied to a fluid at a point is transmitted to all other points of the fluid without loss. Obviously this state of affairs is not true in any real liquid, although it may be approached when short wide flow channels are involved.

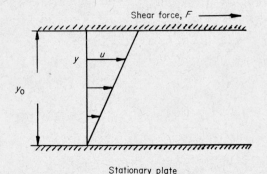

FIG. 2.4. Viscous flow of an ideal (Newtonian) fluid. Velocity of plate $= u_0$; area $= A$; shear stress $\tau = F/A$; shear rate $\gamma = du/dy$.

2.4.2 The Newtonian fluid

We can best describe the stress deformation behaviour of this example of an ideal fluid by considering two parallel plates of very large area A separated at a small distance y_0 by the fluid (see Fig. 2.4). A shear force F is applied to the top plate, producing a shear stress τ equal to F/A. The top plate moves with a uniform velocity u_0 with respect to the bottom plate. The arrows in Fig. 2.4 indicate the velocity of the various layers of fluid relative to the velocities of the two plates. The fluid between the plates because of its viscosity flows in a laminar manner in which the velocity u at a distance y from the stationary plate is given by:

$$\frac{u}{u_0} = \frac{y}{y_0}$$

The viscous drag between neighbouring layers of fluid and between the fluid and the plates causes the relative motion u_0 of the plates to distribute itself uniformly across the fluid so that the rate of shear du/dy is constant throughout. The viscous drag shows as a force opposing the motion of the moving plate. To keep the plate in motion a tangential force or shear stress τ per unit area must be applied to the plate. At low rates of shear all fluids obey the relationship

$$\tau = \frac{\eta \, du}{dy} \tag{2.9}$$

This is known as Newton's law of viscosity where η is the coefficient of viscosity of the fluid, usually measured in poises ($\eta = 1$ poise when a stress of 1 dyn cm^{-2} produces a shear rate 1 sec^{-1}). Newton's law is a linear law, the strain rate increasing linearly with the stress, and is equivalent to Hooke's law with η equivalent to the shear modulus G. Typical values of η for different materials at or near room temperature are as shown in Table 2.1.

TABLE 2.1

Viscosities of different materials

Material	Viscosity η at 20°C (*poise*)
Air	$1 \cdot 8 \times 10^{-4}$
Pentane (C_5H_{12})	$2 \cdot 5 \times 10^{-3}$
Water	10^{-2}
Treacle	10^3
Window glass (800°C)	10^5
Pitch	10^{10}

Viscosity values are temperature-sensitive, decreasing exponentially with the reciprocal of absolute temperature T:

$$\eta = \eta_0 \exp (Q/RT) \tag{2.10}$$

or

$$\ln \eta = \ln \eta_0 + Q/RT \tag{2.11}$$

If we define fluidity f as

$$f = \frac{1}{\eta} \tag{2.12}$$

we can, by combining eqns. (2.11) and (2.12), show that

$$\log_{10} f = \log_{10} f_0 - Q/2\cdot3RT \qquad (2.13)$$

Both η_0 and f_0 are constants for a given fluid and Q is an activation energy for the viscous shear of the atoms as they pass one another.

Figure 2.5 shows the dependence of fluidity f on temperature for molecularly simple materials.

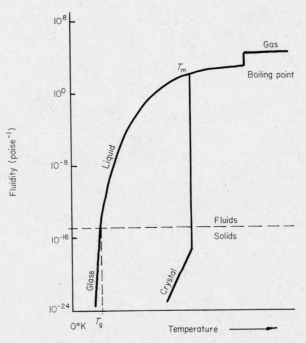

FIG. 2.5. Variation of fluidity with temperature.

On the basis of eqn. (2.13) fluidity f is a diffusion coefficient per unit shear force, $(\text{cm}^2\ \text{sec}^{-1})\ \text{dyn}^{-1}$ and may be compared with other diffusion coefficients in such areas as:

(*a*) *Atomic diffusion*—where D is known as the diffusion coefficient or diffusivity (units $\text{cm}^2\ \text{sec}^{-1}$), and is related to the net flux J of atoms in Fick's first law of diffusion by the following expression:

$$J = -D\frac{\mathrm{d}c}{\mathrm{d}x} \qquad (2.14)$$

where $\mathrm{d}c/\mathrm{d}x$ is the concentration gradient.

(b) *Electrical mobility*—where η the mobility of the charge carriers ($cm^2 \, sec^{-1} \, volt^{-1}$) is related to the conductivity as follows:

$$\sigma = nZq\mu \qquad (2.15)$$

where n is the number of charge carriers, and Zq the charge carried by each.

2.5 FLUID FLOW: NON-IDEAL BEHAVIOUR

As we have already seen η for Newtonian fluids is constant irrespective of the shear stresses involved and by definition is independent of time. Many real materials, particularly polymer melts and solutions and suspensions of particles in liquids, such as PVC pastes, show considerable deviations from Newtonian behaviour. The various types of flow behaviour in fluids can be divided into three broad classes as follows:

(a) Time independent fluids. These are fluids in which the rate of shear at a given point is some function of shearing stress at that point and nothing else. (A Newtonian fluid is in fact the simplest example of this class.)

(b) Time dependent fluids. These are more complex systems in which shear stress/shear rate relationships depend on how the fluid has been sheared and on its previous history.

(c) Elastico-viscous fluids. These are systems which are predominantly viscous but which exhibit partial elastic recovery after deformation. This class could be considered as a special sub-division of (b) but is normally treated separately.

The last two classes are important to us and we shall look at them in more detail later. First of all, it will be helpful to consider the factors which affect the viscosity of materials.

2.6 FACTORS AFFECTING FLUID FLOW

2.6.1 Dependence on temperature

In all liquids, including polymers, the viscosity is very sensitive to changes in temperature. For simple low molecular weight materials the dependence of viscosity on temperature is found experimentally to agree with the formula:

$$\eta = Ac \exp (b/T) \qquad (2.16)$$

where A and b are constants which depend on the nature of the material (Fig. 2.6).

If we consider a simple liquid to consist of a number of roughly spherical molecules, we find that their distribution corresponds quite closely to that in an amorphous solid: that is to say, a more or less regular arrangement containing a certain number of holes or vacant sites. Although there will be a constant change of structure arising from the thermal energy of the system, i.e. molecules will move backwards and forwards into vacancies, these changes will be entirely random in the absence of applied stresses, and there will be no overall movement of the liquid.

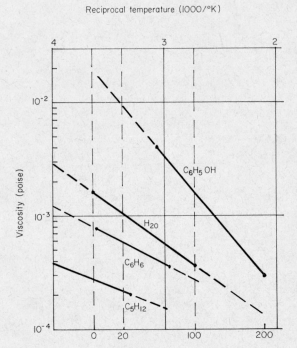

FIG. 2.6. Viscosities for molecules with weak intermolecular bonds (after L. H. Van Vlack).

The application of a shear stress will introduce a directional effect, tending to slide one layer of molecules over another. Thus a molecule will easily be able to jump one way into a vacancy but a jump in the opposite direction will be harder to achieve. This preferential direction of jumping averaged over all the molecules will produce a resultant flow in the direction of the applied stress.

This jumping process requires energy for the molecule to be able to break free from its neighbours. The probability that a molecule will possess this 'activation energy' obviously increases considerably with increasing temperature as we can see from the following example.

The Maxwell–Boltzmann formula gives the chance that a molecule has the necessary thermal activation energy for such a jump, and the number n of successful jumps per second is given by:

$$n = \alpha v \exp(-q/kT) \tag{2.17}$$

where q is the activation energy, v is the number of vibrations per second performed by this molecule and α is an entropy factor connected with the changed vibrational and thermodynamic conditions of the restraining atoms. A typical value for α is about 10.

If $T + \Delta T$ is the temperature at which the rate of jumping is ten times that at T we find:

$$\exp[-q/\{K(T + \Delta T)\}] = 10 \exp(-q/RT) \tag{2.18}$$

whence, by taking logs and approximating:

$$(T + \Delta T)^{-1} \simeq \{1 - (\Delta T/T)\}/T$$

or
$$\frac{\Delta T}{T} = \frac{2 \cdot 3kT}{q} \tag{2.19}$$

Thus, if we assume that $q = 34kT$ (when $T =$ normal ambient temperature), we find that a 20° increase in temperature increases the jumping rate tenfold.

We can now rewrite eqn. (2.16) as follows:

$$\eta = A \exp(E/RT) \tag{2.20}$$

where E is the activation energy and R is the gas constant.

2.6.2 The effect of structure

Besides temperature, one of the main factors affecting the viscosity of amorphous materials is their structure. As with diffusion, small ions and molecules can move past their neighbours with comparative ease and give rise to low viscosities. Other factors, such as molecular shape, polarity and cross-linking, are also important in the viscosity and deformation of amorphous engineering solids.

The relationship between viscosity and structure is conveniently illustrated in Fig. 2.7 for two glasses. Fused silica has a framework structure; since every SiO_4 unit is co-ordinated with four neighbours, the effect is one of cross-linking saturation. The shear stress for flow is very high because primary bonds must be broken. In contrast, a soda glass has a much less rigid structure with fewer tetra-functional units with the result that lower shear stresses are required for flow.

Figure 2.6 also demonstrates that molecular materials with only weak intermolecular bonds, in contrast to those with ionic bonds or covalent bonds, have low viscosities. Of these materials pentane (C_5H_{12}) is lowest because it is non-polar, i.e. it is a symmetrical molecule and has no dipole

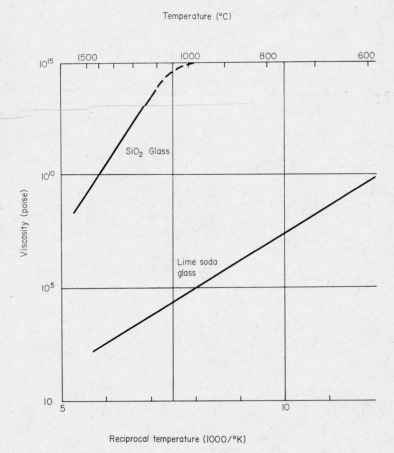

FIG. 2.7. Viscosity *vs* structure. (Courtesy of L. H. Van Vlack.)

moment to create intermolecular attractions. The viscosities for other paraffin compounds with the general formula C_nH_{2n+2} are higher than pentane since, owing to the greater molecular weights, molecular entanglements become increasingly possible.

The extreme example in the paraffin series is polyethylene which, because of its molecular size (n is typically 50 000–100 000) has a viscosity high enough to give it the characteristics of a solid.

The viscosity of water is interesting. Its molecular weight is 18 compared with that of pentane (78). However, the comparatively strong attractive forces (hydrogen bonds) between the molecules mean that a significantly higher shear stress is required to induce viscous flow.

2.6.3 The effect of time

In Section 2.6.1 we considered a simple molecular explanation of flow and its dependence on temperature. In many materials, however, and with

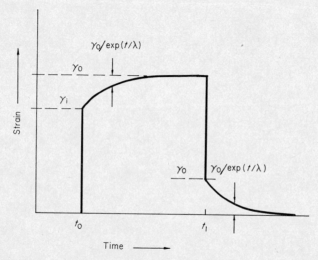

FIG. 2.8. Representation of anelastic strain.

polymers in particular, time is required for these molecular movements. The strain–time curve for this type of process is shown in Fig. 2.8. The time-independent strain γ_i occurs immediately the stress is applied. The additional strain γ_m resulting from the molecular movements into new sites is time dependent, and the total strain $\gamma_i + \gamma_0$ is approached exponentially. When the stress is removed γ_i is recovered immediately, while γ_m (the anelastic strain) relaxes exponentially.

The time dependent strain element behaves logarithmically with respect to time because the rate of atom movements dn/dt is proportional to the number n of atoms remaining behind in the stable sites

$$\lambda(-dn/dt) = n \tag{2.21}$$

Thus

$$\mathrm{d}n/n = -\mathrm{d}t/\lambda$$

and

$$n = n_0 \exp(-t/\lambda) \tag{2.22}$$

where n_0 is the total number of atoms at $t = 0$ which are candidates for removal. λ is the relaxation time, which is defined as the time required to relax all but $1/e$ of the anelastic strain. (At $t = \lambda$, $n/n_0 = 1/e$.)

The number of molecules n_m which have changed places at any time t is

$$n_m = n_0 - n$$
$$= n_0[1 - \exp(-t/\lambda)] \tag{2.23}$$

Since the time-dependent strain is directly proportional to the molecular movements

$$\gamma_m = \gamma_0[1 - \exp(-t/\lambda)] \tag{2.24}$$

we may re-write this

$$\gamma = \frac{\tau}{G}[1 - \exp(-t/\lambda)] \tag{2.25}$$

since by eqn. (2.3) τ/G corresponds to α_0.

The relaxation time λ for viscous flow may be related to viscosity and shear modulus as follows:

$$\lambda = \eta/G$$

from which we may write the anelastic shear strain:

$$\gamma = \frac{\tau}{G}[1 - \exp(-tG/\eta)] \tag{2.26}$$

2.7 DISPLACEMENT MODELS

As we mentioned earlier, it is often convenient—provided we realise their limitations—to represent the various types of deformation response of materials by a series of mechanical analogues.

The ideal (Hookean) solid shows elastic behaviour, as represented by eqns. (2.1) and (2.3). The elastic displacement γ_e is not time dependent, but occurs immediately as shown in Fig. 2.1. The simplest and most convenient method of representing elastic behaviour is by a spring (Fig. 2.9a). With solids of lower elastic modulus the strain/time curve shown in Fig. 2.1 is displaced upwards. The strain is constant but reversible as soon as the stress is removed at t_1.

In the case of the ideal (Newtonian) fluid, which obeys the viscosity relationships:

$$\eta = t\tau/\gamma \qquad (2.27)$$

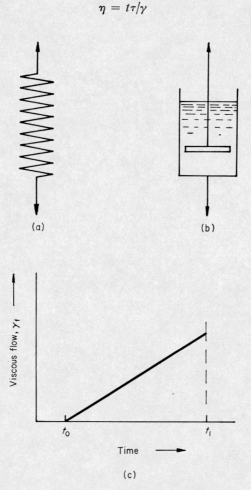

FIG. 2.9. Displacement models for ideal solids and fluids and curve for ideal (Newtonian) fluid. (a) Spring representing Hookean solid. (b) Dashpot representing Newtonian fluid. (c) Displacement *vs* time for a Newtonian fluid.

and eqn. (2.9), we may use a curve of the type shown in Fig. 2.9(c) to show flow displacement γ_f at constant shear stress as a function of time. The mechanical analogue this time is the dashpot (Fig. 2.9(b)), which consists of a loosely fitting piston in a cylinder. Displacement of the piston under an applied force allows fluid to flow round its edges. The displacement takes place more rapidly when the fluid is less viscous. If the force is removed the flow is not reversed; for reversal to occur with a dashpot the force must be reversed.

Most real materials only behave ideally in a very limited number of situations. Generally they show characteristics which are combinations of the two extremes of ideal behaviour. For example, in the case of a silicone polymer, familiar as 'bouncing putty', the material behaves as an elastic solid when rapidly stressed, but if left for several days or weeks it begins to flow and in due course will spread out to form a horizontal surface.

A convenient distinction between solids and liquids is a measure of the rate of flow produced at small applied stresses. This distinction is, however, often neither clear-cut nor indeed particularly meaningful since, as can be seen from Figs. 2.5 and 2.10, it is possible to represent the whole range of

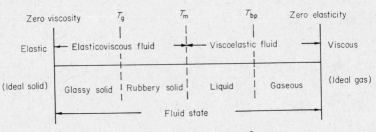

FIG. 2.10. The continuous state of matter.

solid/liquid behaviour as a continuous transition from one to the other. For this reason an arbitrary value for viscosity is often used to define the boundary between solids and liquids. The usual value is 10^{15} poise. At this viscosity a 1-inch cube will bear the weight of a man for a year and deform less than 0·1 inches!

It is sometimes helpful to use the relaxation time λ as a means of distinguishing between solids and liquids. For example, with liquids of high fluidity ($\eta = 1$ poise), λ is about 10^{-11} sec. When $10^{-4} < \lambda < 10^4$ sec, as in waxes, asphalt and putty, both elastic and viscous strain are readily observable. However, when $\lambda > 10^4$ sec the substance behaves as a 'true' solid.

2.7.1 Time dependent deformation

Materials like bouncing putty and asphalt may be represented as a spring and dashpot in series; that is, they have both elastic and viscous components. This representation is known as the Maxwell model and is shown in Fig. 2.11 together with the stress–time diagram associated with this type of material. The Maxwell model is interesting also because it provides a convenient 'lead-in' to the theory of viscoelasticity as well as providing a description of an 'elasticoviscous' material.

Another form of time dependent behaviour is exemplified by the Voigt model, which uses a spring and dashpot in parallel.

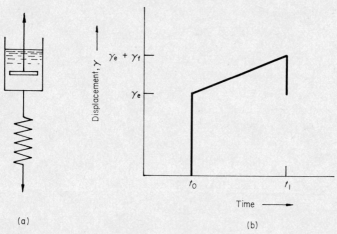

(a)   (b)

FIG. 2.11. Representation of material containing viscous and elastic components. In this (the Maxwell model) the two components are in series. (a) Maxwell model. (b) Stress–time diagram.

The Voigt model together with the associated strain–time curve is shown in Fig. 2.12. With a parallel loading γ_e must always equal γ_f. Initially all the resistance to deformation comes from the viscous component and consequently $d\gamma/dt = \tau/\eta$ (see eqn. (2.21)). Eventually the elastic component carries the full load and the displacement ceases at an elastic strain of τ/G (see eqn. (2.3)).

As we saw in Section 2.6.3 the shear strain may be written:

$$\gamma = \frac{\tau}{G} [1 - \exp(-t/\lambda)] \tag{2.28}$$

or

$$\gamma = \frac{\tau}{G} [1 - \exp(-tG/\eta)] \tag{2.29}$$

2.7.2 The representation of viscoelastic behaviour

In Section 2.2 the problems of realistically portraying viscoelastic behaviour were mentioned together with the behaviour patterns which such a

model must be capable of expressing. The simplest model capable of
illustrating these features is the Maxwell–Voigt spring and dashpot
mechanical analogue, which is shown in Fig. 2.13. The elements are
combined both in series and in parallel, the springs representing the recoverable elastic responses and the dashpots the elements in the structure

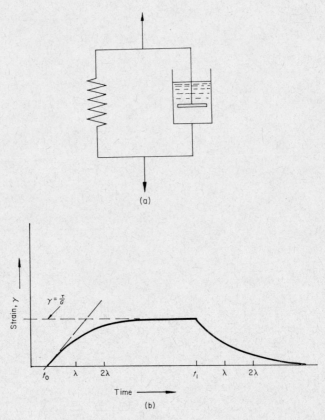

FIG. 2.12. (a) The Voigt model: the elastic and viscous components are in parallel.
(b) The stress–time diagram.

giving rise to viscous drag. The influence of temperature is reflected in the
viscous drag—where the viscosity of the oil in the dashpots decreases with
increasing temperature. The total displacement γ of such a combination
is equal to the sum of the displacements in each element of the series thus:

$$\gamma = \gamma_1 + \gamma_2 + \gamma_3 \tag{2.30}$$

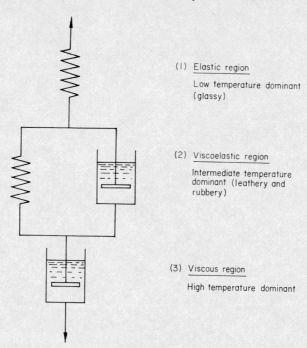

(1) Elastic region

Low temperature dominant (glassy)

(2) Viscoelastic region

Intermediate temperature dominant (leathery and rubbery)

(3) Viscous region

High temperature dominant

FIG. 2.13(a). Maxwell–Voigt model: Maxwell element, Parts 1 and 3; Voigt element, Part 2.

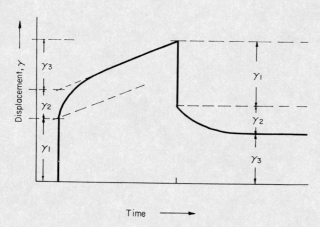

FIG. 2.13(b). Maxwell–Voigt viscoelastic displacement curve.

where the subscripts 1, 2 and 3 refer to the regions of the same number in
Fig. 2.13.

From eqns. (2.3), (2.29) and (2.27) we may write:

$$\gamma_1 = \tau/G \tag{2.31a}$$

$$\gamma_2 = (\tau/G_2)[1 - \exp(-tG_2/\eta_2)] \tag{2.31b}$$

$$\gamma_3 = t\tau/\eta_3 \tag{2.31c}$$

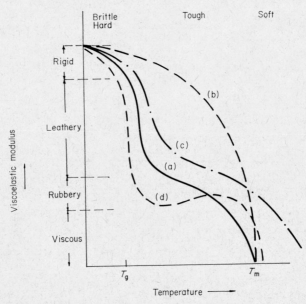

FIG. 2.14. Viscoelastic behaviour of different types of polymer. (a) Amorphous
linear. (b) Crystalline. (c) Cross-linked. (d) Elastomeric.

The importance of the three components in determining the value of γ
(eqn. (2.30)) will depend on the nature of the material and the temperature.
In order to appreciate the extent to which this is so we must now see
how this works in practice. There are many ways in which we could do
this, but a useful method is to categorise polymers into four fairly arbitrary property divisions and discuss them in turn.

These are:

(a) amorphous;
(b) crystalline;
(c) cross-linked;
(d) elastomers.

Their viscoelastic behaviour is shown in Fig. 2.14.

(a) Amorphous polymers

The most familiar examples of this class are polystyrene and polyvinyl chloride. These materials, as we saw in Chapter 1, are devoid of any long range order. In cooling from the liquid to the solid state, the volume contraction is gradual without the sudden jump associated with ordered or crystalline materials (Fig. 2.15). As cooling continues, however, a temperature will be reached below which no further molecular rearrangements are

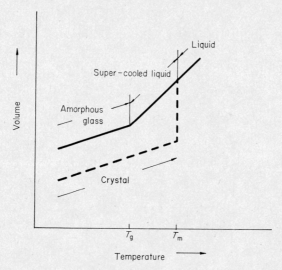

Fig. 2.15. Volume changes of crystalline and amorphous solids. Below the glass transition temperature T_g no further rearrangements occur; additional volume shrinkage is due to reduced thermal vibrations.

possible to achieve more efficient packing. Below this temperature—called the glass transition temperature (T_g)—further cooling only produces a reduction in vibrational amplitude.

The importance of the three components of eqn. (2.30) are shown in Table 2.2. The range $T_g < T < T_m$ is that within which most of the processing operations are carried out. The plastic is rubbery if G_2 is low, or leathery if the spring component becomes stiffer. Within certain limits, as we shall see in a later chapter, it is possible to tailor-make polymers with appropriate properties.

(b) Crystalline polymers

As is the case with metals and other crystalline materials, crystalline polymers show only a gradual decrease in elastic modulus with a sudden fall at T_m—exemplified by curve (b) in Fig. 2.14. The rubbery and leathery

TABLE 2.2

Deformation characteristics of amorphous polymers

Temperature range	Dominant term in eqn. (2.30)	Behaviour
$T < T_g$	γ_1	No facility to adjust to stresses. Material is hard and brittle and behaviour is glass-like
$T_g < T < T_m$	γ_2	η_3 is too high for viscous flow, η_2 is low enough for molecular rearrangement. Behaviour is anelastic
$T > T_m$	γ_3	η_3 is very low, stresses cannot build up; behaves as a true liquid

ranges are absent, but as most crystalline polymers are not fully crystalline, their performance curve is generally intermediate between (a) and (b) of Fig. 2.14. The elastic component γ_1 is the dominant term in eqn. (2.30) and controls deformation up to T_m.

The higher the degree of crystallinity the nearer is the approach to curve (b), and the more suitable will the material be for sustained load-bearing applications.

(c)   Cross-linked polymers

Since cross-linking will obviously reduce the ability to deform under stress, we find that these materials are never able to show high fluidity even in the

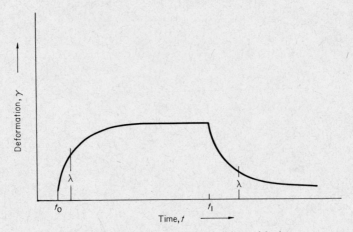

FIG. 2.16. The deformation of rubber with time.

liquid range. Their performance will therefore always lie above that for amorphous polymers (Fig. 2.14(c)).

Below T_g cross-linked polymers are brittle and hard, but in the range $T_g < T < T_m$ limited molecular movement is possible giving rise to a recoverable anelastic displacement. Thus, since both η_2 and η_3 are relatively high, τ_1 and G_1 represent the dominant factors in eqn. (2.30).

(d) Elastomers

These materials are polymers with very low moduli of elasticity (10^6–10^8 N m^{-2}) and are exemplified by the natural and synthetic rubbers. The low value of G is a result of the straightening of molecules which in their lowest energy states show very considerable coiling and kinking.

The values of G_2 and η_2 in eqn. (2.30) are very low, and therefore γ_2 is the dominant term. The behaviour of elastomers with time is shown in Fig. 2.16 and their responses as the temperature is varied are summarised in Table 2.3.

TABLE 2.3
Deformation characteristics of elastomers

Temperature range	Dominant term in eqn. (2.30)	Behaviour
$T < T_g$	γ_1	Material behaves as a hard and brittle solid
$T_g < T < T_m$	γ_2	Immediately above T_g the modulus/T curve drops and then rises as the thermal energy for kinking and coiling is greater. (It takes more stress to produce a given strain.)
$T > T_m$	γ_3	Viscous flow occurs and material behaves as a liquid

2.8 FLOW THROUGH CHANNELS

It has already been pointed out that during conversion polymers are subjected to various intensities of heat and pressure. We shall find, too, that in all extrusion-based processes and in injection moulding, the polymer melt is forced through a series of channels. This will have the effect of causing the polymer chains to uncoil as they are sheared. The entanglements normally present in the material will not only bend to prevent this process taking place, but, once the shearing stresses have ceased to operate, will cause the molecules to retract and attempt to regain their former configurations.

The effect of this behaviour on the operation of the various processing techniques as well as the properties of the product is often considerable, and will be discussed in Chapter 4. We shall, however, consider two general effects here: die swell and unstable flow.

2.8.1   Die swell

When Newtonian liquids and other non-polymeric materials emerge from an orifice into the atmosphere there will be a reduction in the cross-section of the extrudate. Most polymer melts, on the other hand, show an expansion.

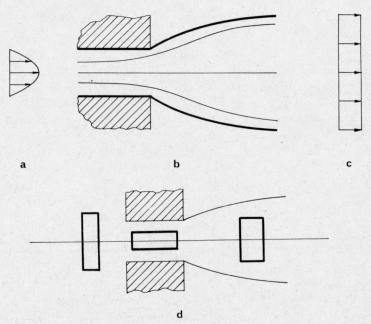

FIG. 2.17. Die swell. (a) Velocity profile upstream in die. (b) Flow lines in polymer emerging from die. (c) Profile downstream of emergent polymer. (d) Illustration of elastic recovery for flow through a short die; elements represent shape of material at various stages during passage.

A considerable amount of work has been carried out on the phenomena of die swell and many theories advanced to explain it. The most generally accepted hypothesis, however, is based on elastic recovery, shown diagrammatically in Fig. 2.17. It is assumed that the molecules are oriented in the die and that after emergence they recoil. Molecules nearest to the die wall are subjected to the greatest shear, so they will tend to shrink to a greater extent than those near the centre, and this effect appears as a convex profile on the emerging extrudate.

As may be expected the amount of die swell is governed by shear rate and pressure and also is strongly influenced by temperature (Figs. 2.18 and 2.19).

2.8.2 Unstable flow

Two phenomena which are seen to occur with elastic fluids are surface roughness or sharkskin, and melt fracture.

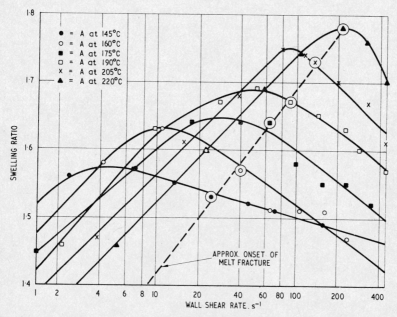

FIG. 2.18. Swelling ratio against shear rate for low density polyethylene. (Courtesy D. L. T. Beynon and B. S. Glyde.)

(a) Sharkskin

This is shown as a lack of surface gloss in the extrudate which appears to be due to the formation of a series of ridges perpendicular to the direction of flow. It is strongly temperature dependent and, like melt fracture (Figs. 2.18 and 2.19) can be delayed by increasing the processing temperature.

This fact, together with its dependence on extrudate velocity, indicates that the changes in acceleration at the surface of the extrudate caused by a rearrangement of the velocity profile throughout the material are sufficient to exceed the tensile strength of the material. Healing of the surface can occur at higher temperatures and lower speeds where viscous flow predominates, but at low temperature and higher speeds elastic behaviour becomes more prominent and breakdown occurs.

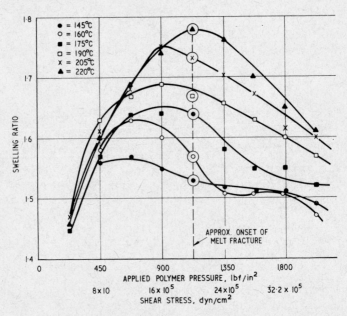

FIG. 2.19. Swelling ratio against pressure for low density polyethylene. (Courtesy D. L. T. Beynon and B. S. Glyde.)

(b)   Melt fracture

This effect is noticed most frequently as the appearance of a distorted or irregularly shaped extrudate varying from a slight waviness to complete break-up. As with other types of melt instability there are differences of opinion as to the causes of the various phenomena. In common with sharkskin the source of the trouble lies in the region of the die wall and seems either to be caused by stick-slip of the melt against the metal surface, or at the region of highest shear in the melt close to the die wall. Figures 2.18 and 2.19 show that the phenomenon is dependent on temperature and pressure. Unlike sharkskin, which appears to be a function of the linear output rate $(q/\pi R^2)$, melt fracture varies with the critical apparent wall shear rate $(4q/\pi R^3)$.

FURTHER READING

Brydson, J. A. (1970). *Flow Properties of Polymer Melts*. Iliffe, London.
Jacobi, H. (1963). *Screw Extrusion of Plastics*, Iliffe, London.
Jenkins, A. D. (ed.) (1967–70). *Progress in Polymer Science*, Vols. 1–3, Pergamon Press, Oxford.
Van Vlack, L. H. (1970). *Materials Science for Engineers*, Addison-Wesley, London.

CHAPTER 3

PRELIMINARY PROCESSING OR COMPOUNDING

3.1 INTRODUCTION

In order to be converted into useful articles, polymers have to undergo a sequence of processing and fabrication stages. The choice of a particular sequence will be dictated by a number of technological and economic considerations. These will include, for example:

(a) the nature of the polymer;
(b) the type of article to be produced, i.e. film, fibres, moulded products, etc.;
(c) the size of the article;
(d) the quantity to be produced.

The manufacturing route, however, may not always entail the direct conversion of the polymer into the finished article, but may start with monomers which are mixed and allowed to polymerise *in situ*. This is particularly the case with thermoset polymers, which are also almost invariably used with fillers or reinforcing agents. Some of these systems are discussed in a later chapter.

As a rule, before passing through such process stages as moulding, extrusion, casting, etc., the polymers themselves are required to undergo preliminary processing stages in which certain additives, ancillary materials and even other polymers are incorporated.

The incorporation of additives can be done at a number of different stages before processing, during the polymerisation reaction, into one of the monomers, or into the polymer itself.

The additives put into polymers may be in the form of gases, liquids or solids, and each will modify the polymer properties in a specific way. Briefly, we may say that the addition of gases produces expanded or foamed materials, while liquids are included as plasticisers, lubricants or stabilisers. Solids, which are available in considerable variety, will affect the resulting material properties according to their own physical characteristics: shape, size, hardness, etc. In all cases, the volume fraction of the

56

dispersed additive will also have an effect on the nature of the product. The precise contribution of different types of solid additives is often difficult to establish, although much work has been done, particularly on fibre-reinforced materials. Very simply we can make this distinction:

(a) Finely divided solid additives generally act as inert fillers and extenders, and in appropriate cases provide protection against degradation.

(b) Fibres are used to strengthen the material, and can be used in forms ranging from single fibres to closely woven mats.

Whatever the other reasons for incorporating additives into a polymer, their function must also be to assist in providing a homogeneous mass of material in a form that is most suitable for processing by the appropriate equipment. The technique used will vary both according to the form of the ingredient and the method selected for fabrication. Generally speaking, however, one of those methods summarised in Table 3.1 will be used.

TABLE 3.1

Techniques used in compounding polymers

System	Conditions	Degree of mixing
Paddle mixers and blenders	Cold or hot	Slight to considerable
Mills—two or more rolls	Cold or hot	Slight to considerable
Internal mixers	Hot	Slight to considerable
Mixer/extruder	Hot	Slight to considerable

The primary reasons for the use of additives or ancillary materials are as follows:

(a) modification of physical, mechanical and electrical properties;
(b) prevention of degradation due to heat or exposure to ultra-violet radiation;
(c) to 'extend' the material and thus provide a cheaper product;
(d) to confer colour, opacity or other visual effects;
(e) as an aid to processing.

Having appreciated why additives are used, we should now consider briefly the three classes of additives; gases, liquids and solids, and the special techniques of mixing and compounding that are used. We are concerned here primarily with methods, but inevitably we must at least touch on properties since the two are so closely inter-related. A more detailed discussion of the properties and applications of this important class of materials in combination, or composites is, however, reserved for Chapter 7.

Polymer Conversion

3.2   THE ADDITION OF GASES: EXPANDED MATERIALS

Thermoplastic foams are generally derived from the heat-softened polymer by the use of a suitable expansion system. Foams made from thermosetting polymers are made from the uncured resin by an expansion system which acts during the curing stage. A variety of such systems is available and their characteristics and applications are summarised in Table 3.2.

TABLE 3.2

Expansion systems for foamed polymers

Expansion system	Method of operation	Application
Air	Air is mixed mechanically into the system which is then cured by heating	Used in manufacture of, e.g. urea formaldehyde foams. Also with thermoplastics in liquid phase bulk polymerisation
Dissolved gas	Gas (usually CO_2) is dissolved in the system under pressure. Foaming occurs when pressure, during the hardening stage, is reduced	Used in production of plasticised PVC foam
Volatile component of mix	A solvent is added to the material which vaporises during the hardening stage	Cellulose acetate foams
Gas evolved during curing process	The gas, which is usually water vapour or CO_2, is formed by chemical reaction in the resin	A typical example is polyurethane
Gas evolved from a blowing agent	The blowing agent is mixed with the polymer. It then decomposes when the material is heated to the appropriate temperature and produces the gas. The system used is chosen so that its decomposition temperature matches the softening point or melting temperature of the polymers	Many applications notably polyethylene and polystyrene. Polymers incorporating systems of this type can be extended and moulded to give foamed products

3.3 THE ADDITION OF LIQUIDS

3.3.1 Plasticisers

There are two main reasons for the incorporation of plasticisers into polymers; firstly, to improve the flow characteristics of the material. This not only allows the polymer to be processed more easily, but also reduces the likelihood of thermal damage since lower temperatures are generally used. Secondly, the use of plasticisers provides a means of controlling the properties of the finished article.

Plasticisers are rarely added to thermosetting materials but are most often used in rubbers and the vinyl and cellulosic classes of thermoplastics. The action of the plasticiser is to decrease the attractive forces between the polymer chains thereby increasing their relative mobility. This can be achieved in two ways; by the use of an insoluble oil which is dispersed

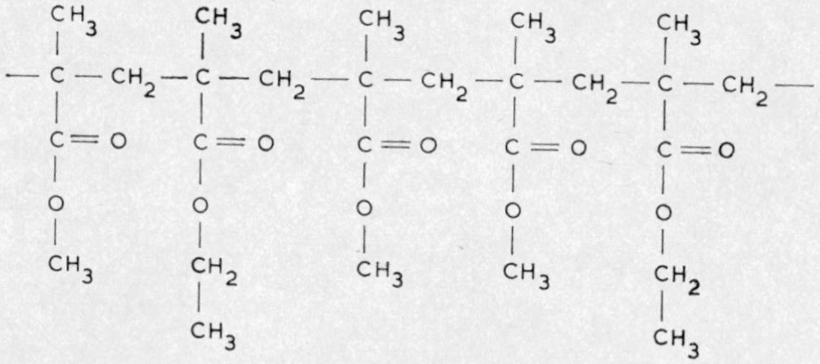

FIG. 3.1. Representation of a methyl/ethyl methacrylate copolymer.

throughout the material in the form of extremely small droplets and which therefore act as 'ball bearings' between the polymer molecules. Alternatively, we can use a non-volatile solvent which produces the effect of a highly concentrated solution with the solvent molecules dispersed throughout the bulk of the polymers. When applied to crystalline materials the solvent is unable to penetrate the crystalline regions and plasticisation occurs by solution of the amorphous areas. These act as extensible links between the crystalline regions.

The incorporation of liquids into the polymers in this way is called external plasticisation but a similar effect can also be achieved by the method of internal plasticisation. Here the polymer chain is modified by the introduction of small amounts of another monomer unit which has a bulky side group. In this way the chains are forced more widely apart and the effect of the intermolecular forces is reduced. A typical example is the copolymerisation of methyl methacrylate with the butyl or ethyl ester. This is shown diagrammatically in Fig. 3.1. A word of caution, however: plasticisation, while improving flow characteristics, will also modify all the other properties of the polymer; there is also the chance of deterioration of the product due to plasticiser migration. In general we can say that, except for its increased flexibility, a plasticised material is inferior in most respects, and therefore the minimum amount of additives should be used to secure the desired performance.

3.3.2 Polymeric materials

(a) Rubber in plastics

In recent years a range of materials has been created by the dispersion of soft rubber particles in a rigid matrix. This is done in order to improve the impact properties of an otherwise brittle material, such as polystyrene, polypropylene or methacrylate polymers.

These compounded materials as well as being useful polymers in their own right can also be used as flow modifiers and process aids in, for example, PVC. Thus we find that the art of blending covers not only the properties of the finished product but has an important role to play in the processing stages.

TABLE 3.3

Methods of preparing rubber-modified polymers

Method	*Features*
1. Physical blending	(a) Matrix and rubber mixed above softening point of the matrix in Banbury mixer or on hot roll mill; or (b) Rubber latex mixed with polymer (matrix) latex, coagulated and coagulum dried
2. Bulk interpolymerisation	Rubber dissolved in matrix monomer. During polymerisation cross-linking of the rubber and grafting of the matrix to the rubber molecules can occur. Used extensively for impact polystyrenes
3. Latex interpolymerisation	Combines features of 1 and 2. Rubber is grafted with matrix molecules by emulsion polymerisation of matrix monomer in rubber latex. Grafted latex is mixed with polymer (matrix) latex, coagulated and coagulum dried. Used, e.g. for ABS polymers

Three methods of incorporation are generally used and the main features of each are summarised in Table 3.3.

(b) Plastics and other materials in rubbers

The technology of rubber compounding and the wide variety of additives which can be incorporated are such that the subject really needs to be treated separately. It is appropriate to do so here as it forms a convenient bridge between the addition of liquids and solids.

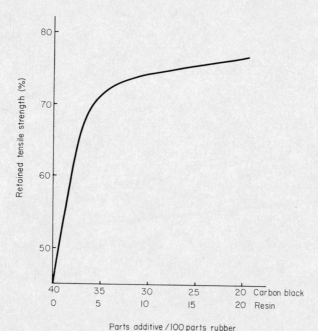

FIG. 3.2. Aging of nitrile rubbers with different loadings of carbon black and thermoset resin.

The methods used in mixing depend on the nature of the additives as well as that of the rubber, and often more than one method is used. Generally speaking, however, one or more of the techniques listed in Table 3.1 are chosen.

The properties of most rubbers, either natural or synthetic, are improved by the addition of carbon black. Further improvements are also obtained by the incorporation of such polyfunctional monomers as methacrylates or thermosetting resins, e.g. of the phenolic type. In the uncured state the additives function as processing aids during compounding and subsequent moulding, and they also enhance the stiffness of the product. Under the influence of heat the resins harden in the presence of aldehyde donors such as hexamine. Thus, by varying the ratio of resin to rubber we can achieve a variety of products ranging from soft vulcanizates to hard ebonite-like materials.

Physical blends of two or more different rubbers with selected characteristics as well as blends of rubbers with, for example, PVC are capable of giving materials with a spread of useful properties.

The interaction between different additives, the presence or absence of carbon black, the amount of additive present, are all factors—sometimes competing—which influence properties.

For instance, the presence of carbon black usually causes an increase in viscosity, while the uncured thermoset resin gives greater ease of processing and a reduction in viscosity.

Figures 3.2, 3.3 and 3.4 give examples of the effect of additives on the aging characteristics and hardness of nitrile, styrene/butadiene (SB) and ethylene/propylene (EP) rubbers.

The method of compounding will also have some influence on the properties of the product. For example, a blend of nitrile rubber and PVC may be produced either by co-precipitation or by mechanical mixing. Since the degree of dispersion and particle sizes is different in each case, the products will exhibit differing physical properties.

3.3.3 Stabilisers and other additives

When exposed to the atmosphere, sun, ultra-violet light and heat, most organic materials will suffer some degradation, and for this reason the appropriate stabilisers, antioxidants and, when required, suitable antistatic treatments, are incorporated.

3.4 THE ADDITION OF SOLIDS

The principle of using different classes of solid materials in combination provides us with a range of materials of considerable engineering importance. These are discussed in more detail later, not only because they are valuable materials in their own right, but because they present some interesting examples of polymer technology, and in some cases need special processing techniques.

3.4.1 Particles

These are used both in thermosetting and thermoplastic materials and are put in for a wide variety of reasons, as summarised in Table 3.4.

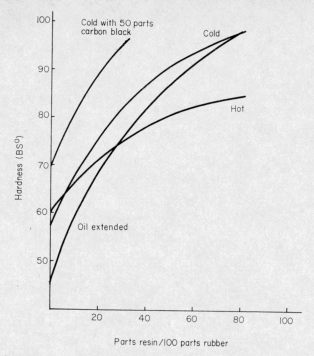

Fig. 3.3. Effect on hardness of the addition of a thermosetting resin to SBR.

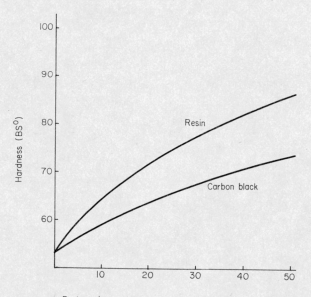

Fig. 3.4. Effect on hardness of the addition of a thermosetting resin or extra carbon black on EP rubber containing 25 parts black/100 parts rubber.

TABLE 3.4

Some powder/polymer systems

Type of additive	Class of polymer	Purpose
Fillers: wood flour, cork dust, paper pulp, asbestos, metal oxides, carbon black, graphite etc.	Usually thermoset, particularly phenolics and aminoplastics	(1) Reduction of brittleness (2) Cheaper product by acting as extenders (3) Modification of other physical properties
Pigments and dyes	Both thermosetting and thermoplastic polymers	(1) Protection of material (2) Protection of contents of pack (3) Decoration
Metallic	Thermosetting polymers	Usually in cold-curing systems for: (1) Electrical and magnetic applications (2) Toolmaking (used for modifying heat transfer characteristics) (3) Decoration
	Thermoplastic polymers	(1) Protection (e.g. lead-filled PVC as a radiation screen) (2) Modification of thermal electrical and magnetic properties (3) Decoration

3.4.2 Fibres

The incorporation of fibres into polymers produces, as we have seen, a range of materials of increased mechanical strength which are called 'reinforced plastics'. Their importance in engineering and many other fields is continually increasing, and in Table 3.5 is given a summary of the materials and techniques generally used.

3.5 TECHNIQUES USED IN MIXING AND COMPOUNDING

3.5.1 Mixing

It is apparent from what has already been said that one of the main requirements of a mixing process is the provision of a product in which the

TABLE 3.5

Some fibre/polymer systems

Fibre type	Form in which added	Effect of addition
Flexible		
Cotton	Cotton linters, or macerated waste cloth, chopped cloth, yarn or cord	Increase in impact resistance, some increase in tensile strength and stiffness
Synthetic textile fibres	Not widely used except in amino resins	Improvement in mechanical strength
Rigid		
Asbestos	Bundles of fibres (commercial fibres are usually bonded with an amorphous cement), non-woven mats	Considerable increase in Young's modulus, good tensile strength, particularly with parallel alignment
Glass	Continuous filaments, chopped strand, yarns, tapes, cords, fabrics, mats	Considerable increase in Young's modulus and tensile strength. Capable of high temperature use in thermally stable polymers
Metallic, inorganic, and organic high modulus fibres	Generally as chopped fibres, sometimes as continuous filaments	Considerable increase in Young's modulus and tensile strength. Capable of high temperature use in thermally stable polymers

components are uniformly dispersed, and that this should be able to be carried out in a reasonably short time.

Mixing has been subjected to much theoretical study, some of which has been backed by experimental work. Generally, however, the criterion for 'perfect' mixing in most cases is in the visual appearance of the mix and whether or not it gives a homogeneous product after subsequent processing.

The mixing process is usually carried out in one of two ways, although before the final blend is achieved it is not unusual for a number of stages to be used often incorporating both methods.

The two methods can simply be described as follows:

(a) Bulk dispersion

Here, the components are simply placed in the mixer in the appropriate proportions and intermingled. The process is most efficiently used when the components are of similar particle size and density. Where liquid and sticky additives are concerned it is often best to introduce them as a fine dispersion into the solid components while these are being agitated in the mixer.

(b) Masterbatching

The additives are incorporated in relatively high concentrations to small amounts of polymer. This makes the problem of efficient dispersion much easier, and the resulting concentrate is then mixed with virgin polymer.

(*i*) *Tumble mixers.*   The simplest form of mixing machine is the tumble mixer in which the ingredients are placed in a drum. The drum is then rotated end over end until the contents are thoroughly mixed. This technique is generally used for providing blends of different batches of polymer or for dispersing a masterbatch in natural material. Several developments have been introduced—particularly in the larger mixers—including internal baffles, multiple chambers and asymmetrical rotation.

(*ii*) *Blade mixers.*   This class of mixer can conveniently be subdivided into two types:

(a) paddle and Z- or Σ-blade mixers;
(b) ribbon blenders.

The essential difference between the two types is that in paddle and blade mixers the mixing elements are generally separate, although they may intermesh, and often the shafts on which the blades are mounted describe quite complex paths within the mixing chamber, which itself may also rotate.

In the case of the Z-blade and Σ-blade mixers the rotors usually operate in two interconnecting mixing chambers. Rotating often at different speeds they provide a complex flow pattern wherein the material being mixed is at times lifted and folded and at times subjected to a steady kneading action.

The ribbon blender is a development of the Archimedean screw in that the blades are usually in the form of strips mounted round a shaft rotating in a trough-shaped mixing chamber. Very little heat is generated by the screw in a ribbon blender and temperature control is achieved by the use of a heated or cooled jacket round the mixing chamber.

(*iii*) *Air or fluid mixers.*   Rapid mixing and efficient dispersion can be achieved by these machines. Generally they operate on the principle of throwing the components to be mixed towards the walls of the mixing chamber and allowing them to return downwards through the centre. The movement of the components is carried out either by controlled jets of air—the fluidised bed principle—or by the use of a rotor in the base of the chamber. Some degree—often very considerable—of frictional heating is usually developed in fluid mixers. This heat is used to cut down the mixing cycle. However, since controlled cooling is virtually impossible to achieve, it is quite often the practice to transfer the mixed material, once it has reached the required state, to a low speed cooling mixer to prevent further agglomeration.

(*iv*) *Miscellaneous methods.* This includes methods used principally for mixing PVC compositions although a number of other techniques are used for specific purposes, such as colloid mills, disc mills and pin mills. All are used where it is necessary to produce intense frictional work on small batches of material, and are effective in dispersing pigments, stabilisers and other additives where the degree of dispersion is critical. Dispersion of pigments in soft polymers is also achieved with an instrument called a muller. This employs a number of rollers which rotate either about horizontal or vertical axes in a cylindrical working chamber. Pigments themselves are often ground in ball mills.

3.5.2 Mills, internal mixers and extruders

It was shown in Table 3.1 that the different methods of mixing each impose their own pattern and degree of severity on the matrix material and additives. Also, in order to achieve the correct properties in the material before conversion by extrusion, moulding, etc., it is often necessary to use several different methods during the preliminary processing sequence. Furthermore, different sequences can be used to produce the same end result. As an illustration of this an additive such as an antistatic agent is mixed at high concentration into a masterbatch, which is then mixed with virgin polymer. Alternatively, the additive could be directed into the mass of the polymer. The nature of the ingredients and their proportions, however, will usually dictate the method.

As an illustration of the variety of methods used in sequence we can take the mixing of natural and synthetic rubbers. The various mixing sequences which are currently available are shown in Fig. 3.5. As can be seen there are many variants in the three basic stages shown, viz.: (a) mastication; (b) two-stage mixing—masterbatching and final mix; (c) single-stage mixing. The choice of system will depend upon the type of material to be processed, the number of compounds being mixed as well as storage and labour considerations.

68   *Polymer Conversion*

(*a*) Mastication

The material is first processed in an internal mixer of the type shown in Fig. 3.6 after which it is passed to one of four secondary mixing systems:

  (i) a two roll mill;
 (ii) an extruder;
(iii) two mills in series;
(iv) an extruder and a mill.

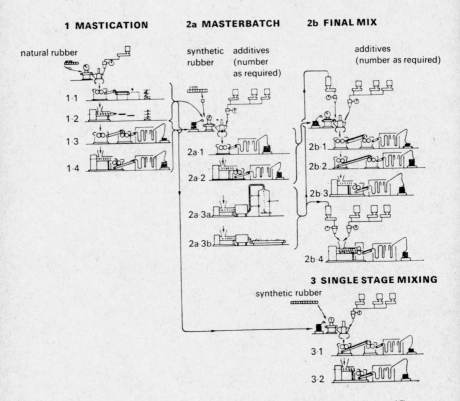

Fig. 3.5. Schematic representation of a number of mixing systems. (Courtesy P. Whitaker, Francis Shaw & Co. Ltd.)

Methods (i) and (ii) tend to be used where low outputs can be tolerated since there is a fair degree of product handling, whereas methods (iii) and (iv) can use automatic feeding to the next stage.

FIG. 3.6. An internal mixer with a batch capacity of 190 litres—the Francis Shaw K7 Unit Drive Intermix.

(*b*) *Two-stage mixing*

This sequence also starts with an internal mixer, a technique incidentally used currently with most PVC formulations, afterwards passing through one of the following:

   (i) a two roll mill;
   (ii) an extruder and mill;
   (iii) an extruder and pelletiser.

In the case of the final mix, the material is passed from the internal mixer through the usual combination of mills and/or extruder. Alternatively, a continuous mixer such as the Transfer mix can be used, but only if the masterbatching unit can provide feed in the form of pellets.

There are advantages and disadvantages associated with the methods described, and to these must be added economic factors such as depreciation and labour costs. In Fig. 3.7 are given details of two modern systems whose operating costs are compared in Table 3.6. It will be seen that taking

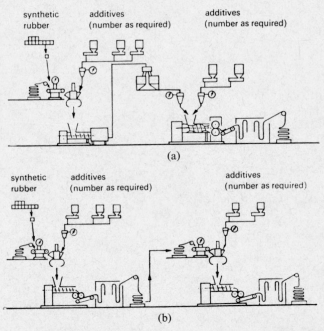

Fig. 3.7. A comparison between two mixing systems. (a) Intermix/continuous mixer system. (b) Intermix/intermix system. (Courtesy P. Whitaker, Francis Shaw & Co. Ltd.)

depreciation and labour into account the twin intermix system is cheaper than the intermix continuous mixer systems. Figure 3.8 compares the annual mixing costs of these systems together with those for an internal mixer with a batch capacity of 640 litres instead of 190 litres. Savings (at 1970 prices) of between £10 000 and £50 000 p.a. could result from using the twin intermix system as opposed to the intermix continuous mixer system. Also at large quantities (23 000 kg/hr) the larger machine provides savings of the order of about £80 000 p.a. Against this, of course, must be offset the increased capital costs and power consumption.

TABLE 3.6

Capital and operating costs of two mixing systems. (Courtesy P. Whitaker, Francis Shaw & Co. Ltd.)

	Intermix continuous mixer system	*Intermix/intermix system*
Output kg hr^{-1}	8 600	8 600
No. of men associated with process	4	4
Capital cost	£410 000	£330 000
Depreciation p.a.	56 000	45 000
Labour cost (75p hr^{-1})	21 000	26 000
Total operating cost p.a. (excluding fuel)	77 000	71 000
Comparative cost per 1 000 kg hr^{-1} output	9 000	8 300
Comparative cost (new pence per kg)	0·129	0·117
Estimated fuel costs	0·083	0·083
Total mixing cost (new pence per kg)	0·212	0·200

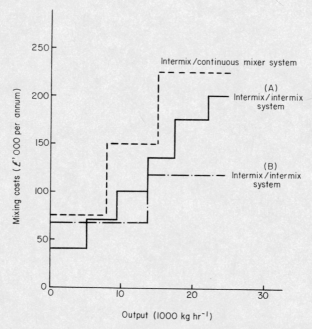

FIG. 3.8. Mixing costs *vs* output for different systems. The intermix used in system (A) has a batch capacity of 190 litres; that in system (B) 640 litres. (After P. Whitaker, Francis Shaw & Co. Ltd.)

Roll mills

Compounding mills, which may be either heated or unheated, are produced with a variety of roll arrangements. Machines with up to five rolls are sometimes used but the most normal is the triple-roll system. Operating without heating this set-up is generally used in PVC technology for dispersing solid additives in plasticisers.

Heated roll mills are usually two-roll systems (Fig. 3.9(a)) in which one roll rotates in fixed bearings. The other can be adjusted with reference to the fixed roll by means of movable bearings operated by adjusting screws.

(a)

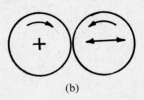

(b)

FIG. 3.9. Details of two-roll mill construction. (a) A 'twin' two-roll mill. Both sets of rolls are driven by the same motor. (b) Arrangement of rolls in a typical two-roll mill; one is fixed, the other movable. (Courtesy Francis Shaw & Co. Ltd.)

The mill performs the compounding operation as a result of the interaction between the surfaces of the rolls through the material in the gap or nip between them. To achieve this, the rolls rotate in opposite directions so as to draw the material downwards into the nip (Fig. 3.9(b)). To provide increased compounding action, it is customary to use different speeds for each roll, usually up to a maximum ratio of 1:1·4.

The working temperature of the rolls depends on the materials being compounded and is kept constant by circulating steam, water or other heat-exchange fluids through passages drilled in the rolls themselves.

In operation, material is fed into the nip and gradually forms a continuous crepe round the front or working roll, which generally operates at a temperature about 3–5°C higher than the back roll. The outputs from two-roll mills are rather low and the labour costs are relatively high. Thus, in production, they are rarely used as the sole means of compounding. Used in conjunction with an internal mixer or extruder they can operate continuously with little need for supervision.

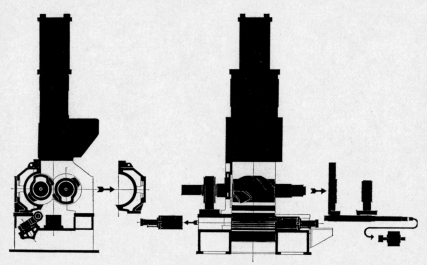

FIG. 3.10. Illustration of the mixing unit of an internal mixer. (Courtesy Francis Shaw & Co. Ltd.)

Internal mixers

These machines are similar to Z-blade or Σ-mixers in that the rotors operate in a closed double mixing chamber (Fig. 3.10). Their mixing action is extremely vigorous due to the interlocking projections on the rotors, which are illustrated in Fig. 3.10. The action, in fact, is very similar to that achieved by a two-roll mill, which has always been recognised as an excellent mixing machine. Its excellence is due to the large area of cooled metal surface in contact with the mix, combined with low friction ratio and a narrow nip. This large surface area together with the screw action caused by the helical projections (nogs) on the rotors combined with the small friction ratio in the nip between the rotors retains the characteristics of the two-roll mill.

It will be seen from Fig. 3.10 that the charging chute of the machine is closed by means of an hydraulically operated ram. The application of pressure on the charge during compounding not only reduces the cycle time, but also improves the consistency of the product (Fig. 3.11). This reduction in cycle time is important for two reasons: (a) it minimises the risk of damage to the materials being compounded by prolonged exposure to heat and mechanical working, and (b) the 'dead time' in the cycle is reduced. The layout of a typical production line based on an internal mixer is shown in Fig. 3.12.

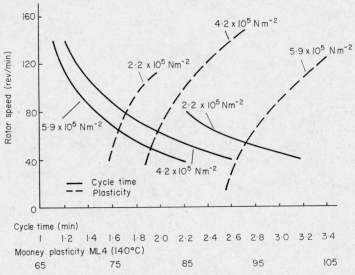

FIG. 3.11. Effect of applied pressure on cycle times and consistency of polymer mixes at a range of rotor speeds.

Extruders

The methods of compounding so far discussed—mills and internal mixers—although capable of providing a mix of the right quality for further processing suffer from the disadvantage that they are batch processes. Apart from the inconvenience associated with intermittent processes, there is the chance of batch-to-batch variation. In order to provide a continuous supply of melt it is necessary to use a form of screw extruder. Although at first sight a conventional single or multi-screw extruder would appear to be capable of performing the task adequately (see Chapter 4), it has been found almost impossible to match the outputs and quality obtained with internal mixers. The problem is to introduce sufficient shearing action

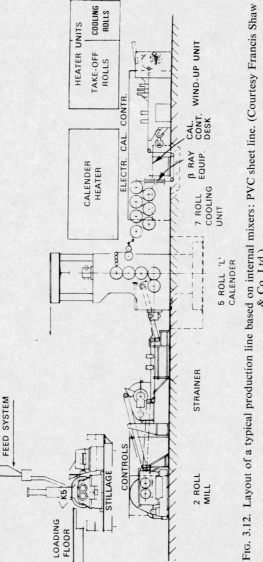

Fig. 3.12. Layout of a typical production line based on internal mixers: PVC sheet line. (Courtesy Francis Shaw & Co. Ltd.)

during the passage of the melt down the extruder barrel. To begin with, this was attempted by fitting ancillary attachments which provided intense localised shear in a direction perpendicular to the normal laminar flow of the melt. Generally, however, better results are obtained with machines specially designed for the purpose and a considerable variety of compounding extruders is now available.

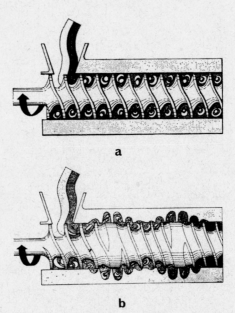

a

b

FIG. 3.13. A schematic representation of the Transfermix, a machine for compounding elastomeric materials. (a) The operation of a conventional rubber extruder. (b) The principle of the Transfermix. (Courtesy C. M. Parshall and A. J. Saulino—Uniroyal Inc.)

One of the most widely used compounding extruders is the Buss Ko-Kneader. It uses a screw with slotted flights which intermesh with a series of pegs in the barrel wall. Superimposed on its normal rotation, the screw is given a reciprocating movement. This breaks up the flow pattern and produces the necessary homogenisation relatively gradually during the passage of the material down the barrel.

For rubber compounding a relatively simple machine called the 'Transfermix' uses a specially profiled screw and barrel (Fig. 3.13), which forces the material sequentially into grooves on the screw and in the barrel and so achieves a controlled pattern of shearing operations. The principle is not dissimilar to that used in the Maillefer screw (see Chapter 4).

The homogenising action of twin-screw extruders is often greater than can be obtained with a single screw, and twin-screw machines are now almost invariably used with unplasticised PVC. Some of the variations available are discussed in Chapter 4.

FURTHER READING

Matthews, G. A. R. (1972). *Vinyl and Allied Polymers*, Iliffe, London.
Miles, D. C., and Briston, J. H. (1965). *Polymer Technology*, Butterworths, London.
Simonds, H. R. (1964). *The Encyclopedia of Plastics Equipment*, Reinhold, New York.

CHAPTER 4

PRIMARY PROCESSING METHODS: CASTING AND MOULDING

4.1 INTRODUCTION

The principal factors which influence the conversion of polymeric materials into the finished article were briefly outlined at the start of the previous chapter, and we shall now discuss the processes themselves. Before doing this, however, there are some points we should consider about processing in order to make it easier to appreciate that the development of processes and materials are closely associated, but not always in step. In the first place, the rapidly increasing range of materials means that the polymer technologist is constantly having to extend the capabilities of the processes which are available to him. This extension process may follow a number of different paths, for example, the use of either extruders or internal mixers to provide feedstock for calenders. Another example is the use of extrusion and injection moulding techniques for processing thermosetting and expanded materials. Secondly, there is the development of new techniques in order to achieve a specific end product. Examples are the adoption of metallurgical forging methods for the solid-phase forming of components, and the use of two injection systems feeding into the same mould (the I.C.I. sandwich moulding process) to produce articles with different material characteristics at the surface and in the interior (Fig. 4.1).

Improvements in both processes and materials have also occurred as a result of the need to increase output as well as the quality of the product. This progress is particularly apparent in moulding (see Section 4.3), where we see the technique of compression moulding leading into transfer moulding, and thence into injection moulding.

In order to reduce this immense and complex subject to manageable proportions, we shall consider first the 'primary processing techniques', where the fabrication of a product is achieved in one operation—which may nevertheless comprise several stages. After this we shall deal with the 'secondary processing techniques'. These are the fabrication techniques, such as thermoforming, which transform a product—sheet or film, for example—into the finished article.

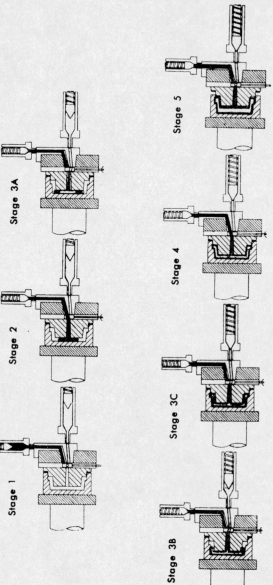

Fig. 4.1. Sequence of operation in the I.C.I. sandwich moulding process.

Although it is convenient to make this rather arbitrary division it must be remembered that primary and secondary processes are often linked in one production line. Also material and process variables interact at all stages of the process and profoundly affect the properties of the product (Fig. 4.2). In addition we must bear in mind what happens to the polymer

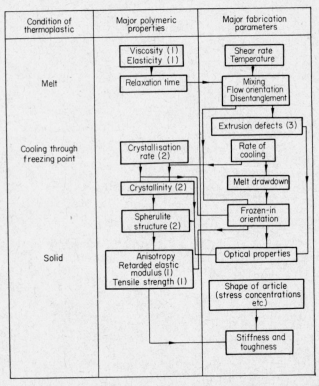

FIG. 4.2. The effect of fabrication on final properties. (Courtesy P. L. Clegg, I.C.I. Ltd.) *Notes:* (1) Time-dependent properties. (2) For those thermoplastics that are partially crystalline. (3) May affect the surface finish of the article and thus its appearance optically.

in the majority of the conversion processes. There are three basic stages: (a) the input of heat to soften the polymer; (b) forming; and (c) removal of heat. All polymers are bad conductors of heat and therefore are susceptible to over-heating. Thus, if the polymer is exposed to excessive temperatures, or to prolonged residence times in the machines, thermal degradation will take place leading to chain scission and reductions in molecular weight and melt viscosity, sometimes continuing to the point of depoly-merisation. The sensitivity of different materials varies very considerably; polyethylene under abnormal conditions still only degrades relatively slowly, polypropylene and polystyrene more rapidly, and PVC does so with considerable ease. Some typical behaviour characteristics for different materials are shown in Fig. 4.3.

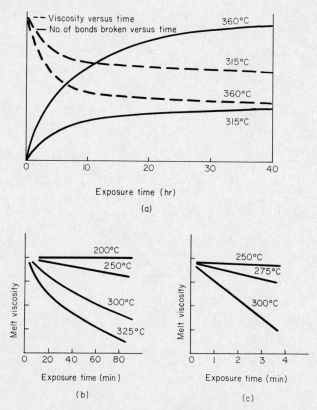

FIG. 4.3. Thermal stability of polymers at various melt temperatures. (a)–(b) Poly-styrene. (c) Polypropylene homopolymer.

4.2   CASTING

In the normal sense of the term, casting implies heating the material until it is molten and then pouring it into a mould, where it solidifies. However, few polymeric materials would be able to withstand the temperatures re-quired, or the danger of attack by oxidation or hydrolysis, and several less demanding variants of this technique have been developed. Properly used they are able to compete successfully with other, more rapid processes such as blow moulding, injection moulding and thermoforming. The methods now in use may be described as follows: (a) polymerisation in the mould; (b) slush casting; and (c) powder casting.

4.2.1 Polymerisation in the mould

In this process, the material solidifies in the mould as the polymerisation proceeds, and not as the material cools.

With thermoplastics the appropriate catalyst is added to the monomer, mixture of monomers, or low molecular weight polymer. The process can then be used for the production of tubes, rods, sheets and other articles from, among other materials, certain polyesters, acrylic and styrene polymers.

The technique can also be used with thermosetting materials which are poured into the mould as liquids and allowed to cure either at room temperature, or in an oven.

4.2.2 Slush casting

One of the early casting processes for metals, used particularly for statuary (lead soldiers!), involved inverting the mould a suitable time after casting. The central core of hot, liquid metal drained away leaving a solidified skin next to the interior mould surface. Very little wastage occurred since the metal from the centre of the casting was re-used.

This method is now used, and has been for many years, with dispersions of PVC in plasticiser called 'plastisols'. The material is poured into the mould, which is often rotated in order to ensure a more uniform coating. Heating the mould then causes the polymer to gel into a hollow flexible moulding.

An alternative version of this technique, sometimes known as 'dip moulding', uses a heated male mould which is dipped into the plastisol. It thus acquires a coating which can be stripped off after curing.

These methods are commonly used to produce such items as dolls' heads, bodies and limbs, soft toys, balls, gloves, grips, gaiters, etc. The use of a large number of individual moulds on the same assembly makes possible a relatively high production rate.

4.2.3 Powder casting

Three techniques, or variants of them, account for most of the powder casting or moulding currently carried out and all of them essentially produce objects by melting the polymer, in powder form, against the inside of the mould. The three methods are: (a) the Engel Process; (b) the Heisler Process; and (c) rotational casting.

The first method uses a static mould, the second a centrifugal hot mould, and the third uses a mould which is slowly rotated about two axes, usually at right angles (see Fig. 4.4). The last method is superior to the others in many ways. It leaves little or no scrap; the exact amount of powder for each cycle can be predetermined, thus providing control over the weight and wall thickness of the product. There is virtually no orientation in articles made by rotational casting, therefore they will be free from internal stresses. The flow characteristics (melt index) of the material, which are of considerable importance in such fabrication methods as blow moulding, are not particularly critical. Furthermore, polymers with a wide range of densities can be used, as can mixtures of different materials.

In view of the relatively slow rate of production, rotational casting is unlikely to compete with, for example, blow moulding, which is a process capable of sustained high production. However, the use of multiple mould assemblies improves production speeds considerably. Also, it must be borne in mind that since the fabrication process does not take place under pressure, very light thin shells can be used for the moulds. This incidentally facilitates the rapid transfer of heat in the heating and cooling parts of the casting cycle.

Some bigger articles, notably storage tanks, are ideal candidates for rotational casting, and, due largely to the development of ultra-high molecular weight polyethylenes, the production in West Germany of storage tanks for heating oil by this method is competitive with glass reinforced polyester and coated steel up to 10 000 litres capacity. Rotational casting as a means of fabricating articles from rigid structural foams is relatively new, and two techniques are currently used. These are: (a) sintering or expansion casting; (b) the production of multilayered sandwich systems having a cellular core.

Both systems use a solid chemical blowing agent which is blended in with the polymer before casting commences. In the case of the multi-layered mouldings accurate control over the degree of fusion of the polymers forming consecutive layers is essential to ensure maximum adhesion between layers after expansion. Complete fusion of the polymer does not in fact occur until the final heating cycle, when decomposition of the blowing agent and total fusion occur virtually simultaneously.

This technique is used to produce, among other articles, free-standing liquid containers, small boat hulls, buoys and floats. The sintering method is particularly useful in the manufacture of items of furniture. Large diameter pipes are also made continuously by rotational casting; the supply of polymer being provided by an extruder which delivers the melt directly into the rapidly spinning mould.

A technique very similar to rotational casting is that used for lining

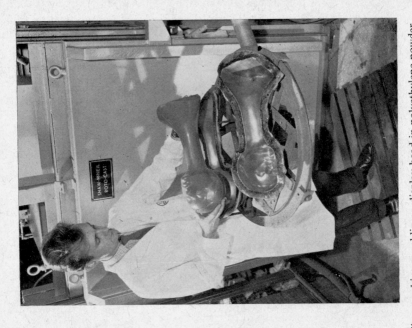

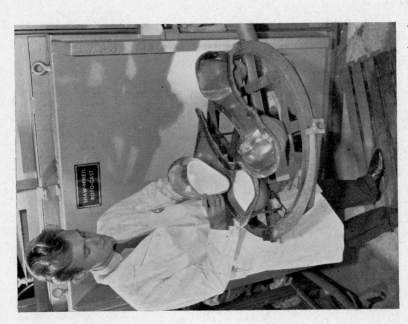

FIG. 4.4. View of the Shaw–McNeil Roto-cast rotational moulding machine moulding a display head in polyethylene powder. (*left*) Filling the mould with powder. (*right*) Extracting the head from the mould. (Courtesy I.C.I. Plastics Division.)

hollow containers, e.g. fire extinguisher cylinders (Fig. 4.5). The pre-heated cylinder is filled with a measured quantity of lining material, which is then sintered and allowed to cool. A typical cycle takes about 8 minutes.

4.2.4 Powder coating

This is a process which may be regarded as being derived from both slush casting and powder casting. An increasing variety of polymeric materials is now applied to substrates for reasons which include: corrosion resistance; electrical and thermal insulation; abrasion resistance; non-stick surfaces; decoration and colour coding. The coating operation is normally carried out by either dipping or spraying.

FIG. 4.5. View showing the operation of a line of machines used for lining fire extinguisher cylinders. (Courtesy Plastic Coatings Ltd.)

Dipping is performed with either plastisols or powders. Where plastisols are used the article to be coated may be dipped cold into a cold thixotropic liquid and then stoved. Alternatively, the article may be heated prior to dipping. In both cases a curing stage is necessary in order to ensure that the coating develops sufficient mechanical strength.

Powder dipping is carried out using the fluidised bed method. The pre-heated article is dipped into a container of cold powder which is held in suspension by upward air currents. This technique has been extended so as to include a range of 'non-coating' applications. One of the first was the use of the suspended particles, in this case small glass beads or 'ballotini', as the heat transfer medium in the continuous vulcanisation of rubber (Fig. 4.6). Another is the replacement of the usual salt bath in the heat treatment of steel articles. Treatment is carried out in beds of mobile particles, such as silica or corundum, which are fluidised by a stream of gas heated either from the combustion of gas/air mixture within the bed itself, or radiant electrical heat.

Spraying is undertaken by two basic methods. In the first the powder is floc-sprayed onto the pre-heated article, which causes them to melt into a continuous film. The second method—electrostatic spraying—charges the powder particles in the spray gun. These are then attracted to the article to be coated, which is earthed. A subsequent sintering stage ensures that a uniform polymer film is produced.

Fig. 4.6. Illustration of the Rapra/Guthrie pressurised fluid bed for the continuous vulcanisation of hose and cable. (Courtesy Rubber & Plastics Research Association, Shawbury, and Guthrie Pullen Ltd., Croydon.)

4.3 MOULDING

4.3.1 Introduction

The moulding process was developed as a means of producing a series of identical articles from a hollow mould. In its simplest form, as already discussed under 'Casting' (Section 4.2), the material is placed in a hollow-mould and allowed to set. The setting process may be accomplished by cooling, evaporation or a chemical reaction, and is often carried out under pressure.

Like many 'universal' processes, this type of simple moulding was used many years ago and finds application in metals and ceramic technology as well as for polymers.

4.3.2 Compression moulding

Although one of the oldest techniques for processing polymers, compression moulding is still regarded as one of the major processing methods. The technique consists of forcing the material into the desired shape using pressures of up to 2×10^8 N m^{-2} (about 30 000 psi). It is applicable to both thermosetting and thermoplastic materials and in its simplest form is referred to as 'cold moulding'. This technique is very cheap but limited in the case of thermosets to materials in the thermoplastic state. The compound mix which is placed in an open cavity may be either in the form of a fine powder or made into a dough by the addition of a suitable solvent. The charge is then compressed cold under high pressure using an hydraulic press. The material softens and flows within the mould cavity and takes up the required shape.

The moulding, which at this stage should have sufficient mechanical strength to withstand handling, is then removed from the mould and placed in an oven where drying, followed possibly by curing or cross-linking, takes place.

More generally, however, the technique of 'hot moulding' is used. Here heat is used as well as pressure and the compounded material is placed in the mould cavity which is heated, after which the pressure is applied. The polymer softens and flows and finally sets to the cross-linked state under the influence of the heat and pressure. The moulding is then ejected and allowed to cool. Temperature, pressure and cure time will depend on the nature of the material and on the size and shape of the moulding.

Typical temperature values for a phenol formaldehyde resin moulding cycle are in the range 150–200°C. Pressures may extend from $1\cdot4$–$2\cdot3 \times 10^7$ N m^{-2} (roughly 2–$3\cdot5 \times 10^3$ psi) and cure times may last from a few seconds to several minutes.

This method, although capable of producing good quality simple mouldings of high dimensional accuracy, is nevertheless inclined to be cumbersome and rather slow. Several extensions to the basic technique have been progressively introduced which are designed to increase the efficiency of the process. The main ones are:

(a) the adoption of multiple cavity moulds;
(b) pre-heating the charge;
(c) the use of moulding materials which have been tabletted or even pre-formed (tabletting consists of pressing the correct amount of

powder into appropriately sized pellets using pressures as high as 1.4×10^8 N m^{-2} (about 20 000 psi);

(d) the use of a screw as a means of preparing the charge.

Preforming extends the technique of tabletting a stage further by producing one large tablet which is close to the size and shape of the finished product. Although by these means an additional process is introduced, the overall cycle time is reduced by shortening the time required to charge the mould.

The use of a screw extruder which feeds the charge—in the form of preheated slugs—more or less directly into the moulds increases output per cavity as much as four times and makes possible the moulding of thick-walled components. This is due to the higher temperatures which can be used and the high plasticity of the material.

Although simple in concept, compression moulds are complex pieces of equipment with devices built in to ensure correct registering of the mould halves, ejection of the finished moulding and the ability to allow the escape of volatiles, such as steam or gases, during the moulding cycle.

There are several different types of compression moulds, which may conveniently be grouped into three categories depending on the means of operation:

(a) hand moulds;
(b) semi-automatic;
(b) automatic moulds.

A further variation also exists, and this is the way in which the 'flash' or excess moulding material is handled by the mould. Moulds are in general designed with either a vertical or horizontal flash and it is usual to separate them as follows: (a) flash type; (b) positive type; (c) semi-positive or semi-flash type. The essential characteristics of these three types are shown in Fig. 4.7.

Compression moulding is principally used for thermosetting materials and for many types of rubbers and, although in principle it can be used with thermoplastics, the need for mould cooling before removal of the product lengthens the cycle and increases the complexity of the process.

For complex components which require delicate inserts the compression moulding process has limitations. Also, since the moulding pressure is transmitted directly onto the material in the mould cavity, the mould components are likely to suffer relatively rapid wear. Further, since the charge is effectively melted *in situ*, there is relatively little movement of the contents—except alas where the inserts are concerned—and this can cause some anisotropy in the moulding which, if severe, may lead to warping in the product. Warping can, however, be reduced by operating the two mould halves at different temperatures, a dodge which incidentally is used to ensure that the moulding is retained in the appropriate mould half.

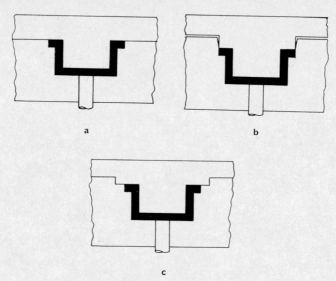

FIG. 4.7. Three types of compression moulds. (a) Flash type. (b) Positive type. (c) Semi-positive type.

4.3.3 Transfer moulding

A further and logical extension to the use of pre-heated and tabletted materials already described involves the actual melting of the charge in a chamber separate from the mould, and then injecting it into the mould itself where it is compressed and cured. The process is shown diagrammatically in Fig. 4.8. Injection through the narrow orifice produces a more uniform temperature distribution throughout the mass of the material, as well as a greater degree of homogenisation. These factors contribute to mouldings—especially in the case of large articles—which are

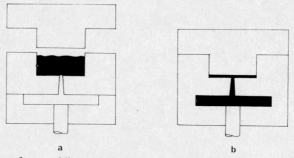

FIG. 4.8. Transfer moulding. (a) Material placed in transfer pot or chamber. (b) Material forced through orifice into the closed mould.

more evenly cured, and which are therefore less likely to distort due to internal stresses. Careful control of temperature is particularly necessary at the preheating stage also, since it is important to avoid curing the polymer to the extent that blockage occurs. A certain amount of early curing can, however, be turned to advantage in reducing cycle times.

A further advantage of transfer moulding is that the mould components

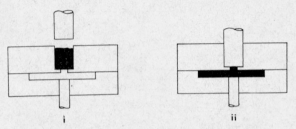

Fɪɢ. 4.9(a). Plunger moulding. (i) Auxiliary ram exerts pressure on material in the chamber. (ii) Material is forced into closed mould.

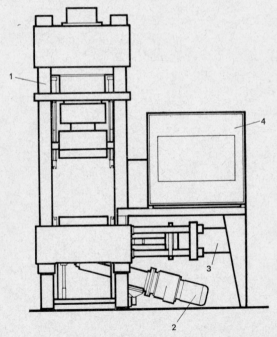

Fɪɢ. 4.9(b). Illustration of the screw-transfer process for rubber moulding. 1. Press. 2. Plasticising unit. 3. Injecting unit. 4. Electrical control unit. (Courtesy Werner & Pfleiderer, Stuttgart.)

do not transmit the pressure directly onto the charge, as is the case with compression moulding, and therefore less wear is likely to occur. It is also possible to mould more intricate shapes, and inserts can more readily be used, since they are not subjected to the high shearing forces encountered in compression moulding. The surface finish on the moulding also tends to be better with this method.

As with compression moulding a number of developments have occurred in the transfer moulding process which are designed to improve the quality of the product and increase output. These include the use of an auxiliary ram or plunger to exert pressure on the material in the chamber. The process—sometimes called plunger moulding—is slightly faster than transfer moulding since the material left in the chamber (the cell), the runners and the sprue, all remain attached to the moulding.

A screw extruder, together with a transfer ram or plunger, gives an assembly (Fig. 4.9) which is particularly well suited to a fully automatic cycle.

4.4 INJECTION MOULDING

4.4.1 Introduction

The technique of injection moulding is basically the logical extension of the transfer moulding process. It is also the simplest of all the moulding processes, since it entails only heating the polymer until it is molten, after which it is forced directly into a cooled mould. A comparison between the various moulding processes is shown in Fig. 4.10.

The early machines which were developed for plastics in the 1930s used a simple plunger or ram to transfer the material into the mould chamber (Fig. 4.11(a)).

Basically we may divide up the injection moulding process into 5 steps:

(a) the mould closes and moves up against the nozzle;
(b) the plunger moves forward and pushes raw material into the cylinder, at the same time injecting plasticised material into the mould;
(c) the plunger remains forward for some time, still maintaining pressure through the nozzle, while the material in the mould is cooling and setting;
(d) the plunger withdraws, the mould remaining closed, and a fresh supply of polymer falls from the feed hopper into the barrel;
(e) the mould moves away from the nozzle, opens, and the moulding is removed.

The use of a reciprocating screw instead of the plunger not only provided a means of heating and plasticising the polymer before injection, but also gave a more homogeneous melt, and, since there were no external charging or metering devices, was able to introduce a much more accurately measured quantity of melt into the mould. It also helped considerably to simplify the process. Figure 4.11(b) shows the essential features of a

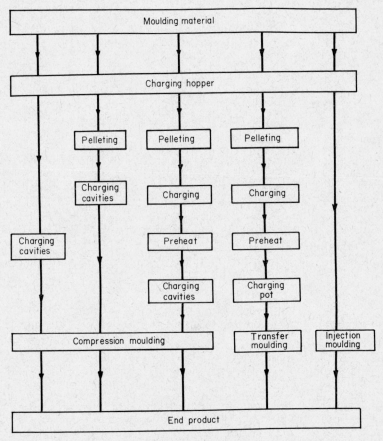

FIG. 4.10. Schematic representation of the steps in moulding processes.

screw-fed injection moulding machine. An illustration of the layout of a typical modern injection moulding machine is shown in Fig. 4.12.

The plunger system, although satisfactory for many purposes, suffers from two main disadvantages: (a) the difficulty in maintaining a consistent shot delivery into the mould; (b) the problem of ensuring adequate plasticising of the material in the barrel without overheating.

Two developments improved the process somewhat: (a) the introduction of a double plunger system, where the second plunger is used only to deliver a measured charge of molten material into the mould; (b) the use of a

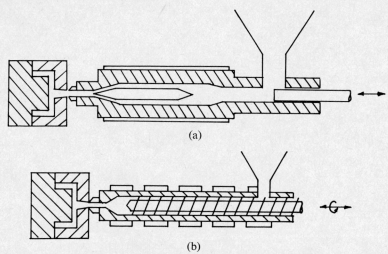

(a)

(b)

Fig. 4.11. Diagrammatic view of conventional injection moulding machines. (a) Ram-fed injection moulding machine. (b) Screw-fed injection moulding machine.

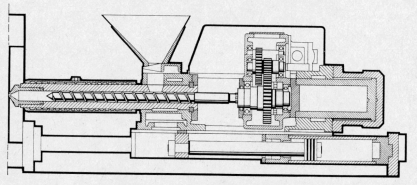

Fig. 4.12. Layout of a typical screw injection moulding machine. (Courtesy Werner & Pfleiderer, Stuttgart.)

screw for plasticising the material followed by a plunger or ram for the actual injection. Both processes, however, require machines of increased complexity and cost. In the case of the screw injection moulding machine the cycle is similar although simplified since several operations occur simultaneously.

4.4.2 The injection moulding cycle

Regardless of the injection system, however, all the operations of the injection moulding sequence take place during what is known as the 'time cycle'. The time cycle is shown in Fig. 4.13, and it is the achievement of the correct proportions of the four interacting elements of the cycle which ensures that mouldings of the highest quality are produced in the shortest possible time.

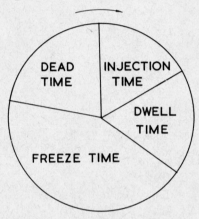

Fig. 4.13. The 'time cycle' in the injection moulding process.

The four elements of the time cycle are described as follows:

(a) Injection—the time taken to fill the mould with material;
(b) Dwell—the time the plunger stays forward maintaining pressure through the nozzle;
(c) Freeze—the time required for the moulding to set sufficiently for safe removal from the mould (the setting process actually starts during the dwell time and continues during the freeze time);
(d) Dead—the time required for the mould to open, the moulding to be removed and the mould to close again.

In addition to the time cycle, successful injection moulding depends on the correct interaction of three other variables:

(i) cylinder temperature;
(ii) injection pressure;
(iii) mould temperature.

We must now consider their effect on the material during the moulding sequence and on the moulded product.

4.4.3 The importance of cylinder temperature

Heat is supplied to the polymer in the barrel in order to convert it into a homogeneous melt of the right viscosity to be transferred from the barrel into the mould. There are several ways in which this heat can be provided:

(a) by means of external heaters;
(b) by mechanical work in the injection screw;
(c) to a lesser extent arising from the flow of the material through the geometry of the system.

We have already seen that polymers, being bad conductors of heat, are susceptible to damage by overheating (Fig. 4.3), and therefore both from a material as well as an economic stand-point the residence time in the barrel must be as short as possible. This period will of course be governed by the complete time cycle, and an unnecessarily high heat input into the polymer will make matters even worse by prolonging the freeze time. Indeed with high speed injection moulding machines—with cycle times of two seconds or less—used for the production of thin-walled drinking cups, the critical factor in raising production speeds is not the time required to fill the mould or remove the product, but the time needed to remove sufficient heat from the moulding to prevent distortion or collapse. The use of heat-exchangers, large diameter mould cooling channels and water under pressures of up to 6×10^5 N m^{-2} have been used to achieve this.

The amount of heat required to bring the material to a suitable viscosity for moulding is a function of several parameters: (a) the specific heat of the polymers; (b) the temperature rise; and (c) with crystalline polymers, the latent heat of fusion. This quantity varies considerably with different materials (see Table 4.1) and therefore the amount of plasticised material a particular machine can deliver will depend upon the polymer used, as of course will the moulding temperature.

TABLE 4.1

Heat required to plasticise a variety of polymers

Polymer	Total heat at moulding temperature ($J kg^{-1} \times 10^5$)
Cellulose acetate butryate	2·8
Polystyrene (general purpose)	2·8
Polyacetal	4·2
Polypropylene	5·8
Polyethylene LD	6–7
Polyethylene HD	7–8

4.4.4 The effect of injection pressure

Injection pressures can vary widely and are dependent on the size and type of machine and the mould design. The aim, when setting the pressure, is to ensure that just enough plasticised material is injected to fill the mould and produce mouldings that are free from sink marks and voids.

As the ram or screw starts the injection stroke, molten polymer passes through the nozzle into the runner system and then fills the mould cavity. During the dwell time (see Fig. 4.13) the injection pressure must be maintained; firstly, to prevent leakage back out of the mould, and secondly, in order to compensate for the contraction of the material in the mould as

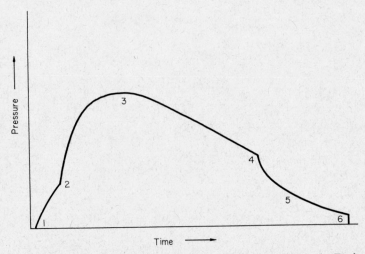

Fig. 4.14. Development of pressure in the cavity during moulding cycle. Explanation of sequence (1–4 plunger moves forward; 4–6 plunger returns): 1–2 cavity filling; 2–3 packing; 3 peak pressure; 4 discharge; 5 gate sealing; 6 residual pressure; 3–6 cooling. (After Rubber and Plastics Research Association.)

it cools. The material in the region of the gate will freeze first, since it is the narrowest part, and after this has occurred the pressure can be released. The development of pressure during the moulding cycle is shown in Fig. 4.14.

Opposing the injection pressure is the force which keeps the mould halves shut. This is called the mould locking force (usually quoted in tons/tonnes) and it must always be greater than the injection pressure developed by the ram or screw.

The ram during its forward stroke encounters a mixture of both solid and plasticised material, therefore there will be a drop in pressure and the pressure at the start of the cycle is higher than the pressure at which the plasticised material is forced into the mould. On the other hand, the screw, as it rotates, is building up a reserve of plasticised material in front of it which gradually forces it back. When it moves forward on the injection stroke it pushes only against plasticised material and therefore the pressure needed for screw injection is less than that needed to operate a ram machine. Thus with a screw machine it is possible to make larger mouldings for the same mould locking force.

4.4.5   The mould temperature

The rate of cooling, as we have already seen (cf. Section 4.4.3), is paramount in determining the rate of production. Just as the plasticisation capacity of the injection moulding machine is dependent on the thermal characteristics of the polymer over the range of operating temperatures, so is the rate of cooling controlled by the enthalpy and heat transfer rate of the material. The rate at which the material cools will also have a profound effect on three characteristics of the moulded article: (a) surface quality; (b) degree of orientation; (c) the amount of crystallinity. The three parameters which we have just considered, namely cylinder temperature, injection pressure and mould temperature, all control the final moulding characteristics. The application of heat and pressure also gives rise to shear and orientation in the polymer melt as it passes through the machine, and we must now look briefly at their effect on the quality of the product.

4.4.6   The effect of shear and orientation

The viscosity of the polymer is reduced by heat and also by shearing. Heat causes a reduction in interchain forces and thus allows chains to move more easily relative to one another. Shearing tends to drag out the normally entangled chains and align them in the direction of flow. The ease with which this process takes place will depend on such polymer characteristics as average molecular weight, molecular weight distribution, degree of branching, presence of additives, etc. Thus the actual shear pressure needed to produce a specific shear rate will vary with different polymers and grades of the same polymer. The higher the shear stress the higher the shear rate produced, and the lower the viscosity of the material. Thus, a high injection rate—which is dependent on the level of the injection pressure—by its effect on the viscosity of the material, helps to fill the mould cavity more easily. The use of increased pressure rather than temperature is a better means of reducing the viscosity since this will not require any lengthening of the moulding cycle by increasing the cooling time.

The effect of shear rate on the viscosity of a selection of moulding materials is shown in Fig. 4.15.

When the hot material enters the cooled mould cavity a complicated molecular re-arrangement takes place, which has a profound effect on the properties of the finished article. The very hot, and therefore relatively little oriented, material from the centre of the melt flow freezes on the cold mould surface. Subsequent layers of material are insulated from the

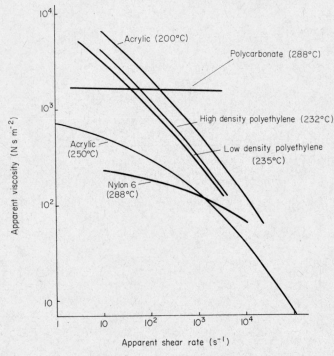

Fig. 4.15. Variation of apparent viscosity with shear rate (after Bernhardt).

mould and cool more slowly. Since they are still subjected to a high injection pressure, the conditions of high shear/low temperature produce a high degree of orientation. When the cavity is full some relaxation of the chains can occur, but this is hampered by (a) the relatively rapid loss of heat to the cold mould, and (b) the sudden increase in viscosity which occurs when shear ceases. Relaxation can occur most at the relatively hot centre of the moulding and the orientation gradient will vary from low at the surface and centre to high in the body of the moulding on either side of the centre line. Anisotropy due to orientation differences will also occur along the line of fill due to differences in cooling rates.

The effect of processing and mould conditions on orientation is shown in Fig. 4.16.

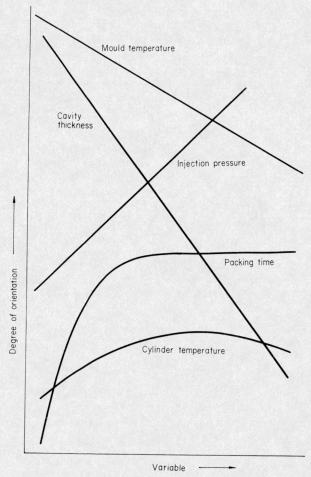

FIG. 4.16. Effect of processing variables on orientation. (Courtesy A. B. Glanvill, Shell Chemical Co. Ltd.)

4.4.7 Mould and runner design

The design and particularly the fabrication of moulds is a complex and intricate matter, and the devices used to produce mouldings having inserts and, for example screw threads, are beyond the scope of this book. Even with a relatively simple mould system, as shown diagrammatically in Fig. 4.17, it is possible to produce components to a high degree of precision (Fig. 4.18).

The size and shape of the passages through which the hot material flows is of importance. For example, in a thick section the velocity will be relatively low, the shear rate will also be low, thus the degree of orientation will be small. The viscosity will, unfortunately, be high. Gate size too is important since, although a small gate offers greater restriction to flow

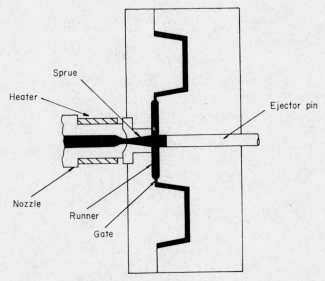

Fig. 4.17. The layout of sprue, runners, gates and cavities in a typical injection moulding tool.

than a large one, this is amply compensated for by the reduction in viscosity caused by the heat generated by passage through the small gate.

During its passage from the cylinder to the cavity the polymer goes through a series of components; first is the sprue bush. The sprue is the material entrance into the runner system from the nozzle of the machine. The sprue length should be kept short and the diameter sufficient to allow adequate material flow, yet permitting the material to 'freeze' as quickly as possible.

Next comes the runner system. The ideal cross-section is circular, but trapezoidal is also adequate; it is convenient to machine and allows easy ejection. The runner should preferably be about 1·5 times the maximum thickness of the component since this ensures that the material in the runner system remains fluid sufficiently long enough for injection pressure to be maintained on the polymer in the cavity to reduce the effects of shrinkage.

The gate is the entry to the mould cavity, and its size and position—which are of considerable importance—depend on many factors which include:

(a) thickness of component;
(b) runner length;
(c) finishing.

FIG. 4.18. Illustration of mould geometry in the manufacture of parts for the Polaroid Swinger camera. (Courtesy Combined Optical Industries Ltd., Slough.)

It is usually advisable to feed into the thickest section of the cavity so that the material impinges on the core or cavity wall. Where multiple cavity moulds are used it is necessary to balance the gate and runner systems to ensure even cavity filling.

Although the system of sprues, runners and gates which has been described is still used in many applications, and has the benefit of cheapness, it suffers from the disadvantage that the material produced in the sprue and runners during each shot must be removed and, where possible, reprocessed.

In an attempt to eliminate these problems several developments have
been introduced in recent years. These include the use of:

(i) hot runners;
(ii) hot nozzles injecting either directly into the cavity or leaving a small
stub sprue—these nozzles may be single or attached to a manifold
(Fig. 4.19);

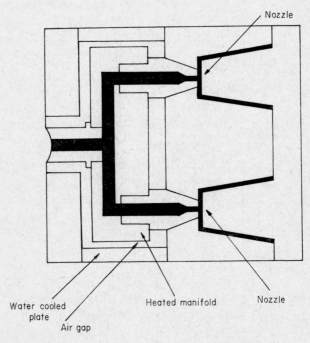

FIG. 4.19. Hot runner mould.

(iii) an insulated runner system (Fig. 4.20)—this may be extended to
include an insulated manifold system to enable edge-feeding of a
mould without offsetting the tool.

(i)   *Hot runner moulding*

In this technique the runners are heated so that the material in the runners
is always molten. When the mould opens the moulding is separated at the
gate and the runners remain full of molten plastic ready for the next
injection.

(ii)   Direct injection nozzles

This technique, illustrated in Fig. 4.19, requires the nozzle to be insulated from the tool. If the area of contact between the nozzle and tool is excessive, the heat loss from the nozzle will cause the melt to chill. Also, the area of the tool adjacent to the nozzle will overheat, resulting in eventual

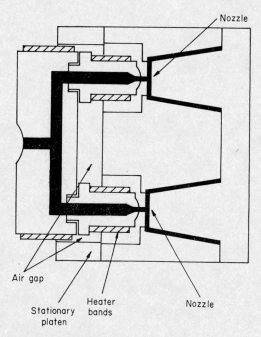

FIG. 4.20. Mould with insulated runner system.

sticking of the material in the cavity due to failure to freeze off. Careful design of the nozzle/tool system is therefore essential, and it is helpful to use cooling water in the area around the nozzle to localise the heat.

(iii)   Insulated runner systems

The basis of this system, illustrated in Fig. 4.20, is the use of a very large runner (up to 2–3 cm in diameter) incorporated in a separate die plate. At the start of the moulding cycle, a layer of material freezes on the runner wall. This insulates the central core of the material which remains liquid, and permits moulding to continue.

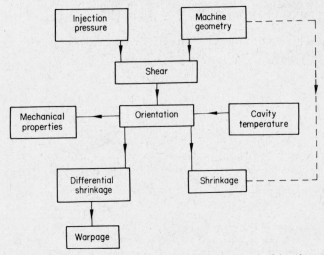

Fig. 4.21. Interaction of processing variables concerned with orientation.

TABLE 4.2

Some factors contributing to common moulding faults

Fault	Barrel temperature	Mould temperature	Injection pressure	Injection time	Dwell time	Cooling time	Mould locking force	Mould venting	Gate design	Ejection system	Feed setting	Water content of polymer
Brittleness	*	*	*								*	
Bubbles			*	*					*			
Burr marks			*					*				
Crazing		*	*								*	*
Distortion	*	*	*	*		*						
Flashing	*		*				*				*	
Flow marks	*								*			
Short mouldings	*	*	*	*					*		*	
Sink marks	*		*		*	*			*		*	
Sticking	*	*	*								*	
Surface gloss	*	*										
Voids	*	*	*		*							*
Weld lines	*	*	*						*			

4.4.8 The interaction of moulding variables and their effect on product quality

We have traced the passage of the material through the internal geometry of the machine and learned what happens to it during the moulding cycle. The influences are complex and interdependent, as can be appreciated from considering just one part of the 'spectrum'—the variables associated with orientation (Fig. 4.21 and Fig. 4.16). Table 4.2 shows some of the more common faults that occur in mouldings and an asterisk indicates the factors which to a greater or lesser degree influence the finished moulding. Since we do appreciate, and measure, the effect—sometimes indirectly —of these variables on the polymer, we can use this knowledge to alter the machine settings and adjust the other moulding parameters in order to obtain good quality mouldings.

4.4.9 Developments in injection moulding techniques

Central to all currently available injection moulding machines is the ram and the screw. Normally either one or the other is used to provide both plasticisation and injection pressure. However, as we saw in Chapter 3 on mixing and compounding, and will again in Chapter 5 on extrusion, the enthusiasts are prone to combine a wide variety of techniques to achieve delivery of a well-homogenised melt. One development, also used in transfer moulding and the moulding of rubbers (cf. Figs. 4.9(a) and (b)), is to use a screw for plasticising together with a plunger-operated injection unit. In the case of the GKN Windsor/Rolinx machine (Fig. 4.22) plasticisation is achieved by using a twin-screw extruder, and the energy for the injection stroke is provided by the use of nitrogen accumulators. The generation of high injection pressures and the power needed to move large masses of machinery rapidly to achieve high platen and injection speeds becomes extremely costly, and the efficiency falls off rapidly as machine size increases. It is claimed that the use of nitrogen accumulators has increased efficiency from 10–15% to at least 35%, and that the motor size has been reduced from over 1000 hp to 250 hp.

The nature of the injection moulding process makes it pre-eminently suitable for the production of large quantities of identical articles; drinking cups and milk crates being two typical large quantity applications.

There are various ways in which large outputs can be achieved, including faster cycling machines and the use of multiple cavity moulds. Each product requires a detailed examination of material, processing and economic factors, but one factor is inescapable; this is the enormous rise in the capital outlay on machinery as the size of the product increases. Figure 4.23 shows the relationship between moulding shot weight, mould locking force and basic machine cost. The spread is due to the availability of different screw and barrel assemblies for each machine. Nevertheless the dotted line, which represents the mean shot weight, agrees closely with the specifications and prices of a number of different manufacturers.

The larger the machine the greater tends to be the inflexibility, and a number of developments have helped to widen the field of injection moulding applications. The first is the use of spark erosion in the manufacture of moulds. This allows the production of relatively complex shapes more rapidly and cheaply than by using the traditional metal working methods. The second is by the introduction of a process, often called 'injection

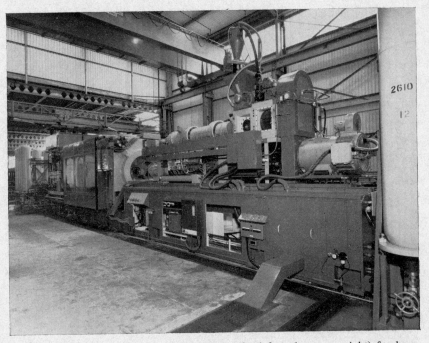

FIG. 4.22. The use of nitrogen accumulators (far left and extreme right) for large injection moulding machines. (Courtesy G.K.N., Windsor.)

stamping', which enables large mouldings to be made on conventional moulding machines. Using normal moulding methods the mould is kept closed under almost constant locking forces during injection, pressing and cooling. Normally the larger the article to be moulded the larger the pressure required to keep the mould shut against the injecting pressure (Fig. 4.23). Using the ratio of flow-length to section thickness, we see that as the projected area of mouldings increases (Table 4.3), the locking force rapidly rises to a very high level. The 'injection stamping' process uses a

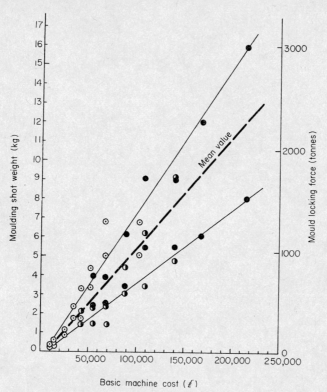

FIG. 4.23. Plot of machine cost *vs* moulding shot weight and mould locking force.

TABLE 4.3

Large area mouldings

Article	Projected area (*m²*)	Ratio of flow length to section thickness	Mould locking force (*tonnes*)
Tray	0·12	90:1	100
Battery box	0·032	130:1	450
Bucket	0·048	265:1	300
Bowl	0·12	170:1	300
Basket	0·12	180:1	450
Dustbin	0·16	300:1	1000
Dome	0·78	310:1	1500

relatively small pressure on the mould which is then allowed to 'breathe' while being filled with the moulding material during the injection stage. The mould is then reclosed quickly and the part is 'stamped' out.

The use of twin injection units also offers potential economies. Although the basic capital cost of one 1200 tonne machine against two 600 tonne machines is about £65 000, compared with £70 000, the increased cost is due mainly to accumulators, hydraulics and co-ordinated electric equipment, and the advantages lie in decreased running costs and considerably increased mobility and moulding flexibility.

Fig. 4.24. An experimental sandwich moulding unit built by Bone Cravens Ltd. (Courtesy I.C.I. Plastics Division.)

Both of these ideas—with slight modifications admittedly—are included in such techniques as the I.C.I. sandwich moulding process (Figs. 4.1 and 4.24). The principle of the process is the injection of two polymer formulations from separate injection units (Fig. 4.24) one after the other into a mould through the same sprue. The mould is then opened a short time after filling it in order to allow the foamable polymer in the core of the moulding to expand, if a foamed sandwich is required. For unexpanded mouldings the foaming agent is omitted.

Using this process a range of sophisticated composites can be made including soft skin/hard centre; hard skin/soft centre; solid skin/solid centre, etc. An example is in the manufacture of a gear wheel. Here a low-friction, hard-wearing material such as p.t.f.e. filled nylon is used for the teeth, and a more rigid material such as glass-reinforced nylon for the centre flange (Fig. 4.25).

We have considered some of the developments which have been and are being used to improve the efficiency and versatility of the injection mould-ing process, and now we must extend our survey to include some of the techniques used with different classes of materials.

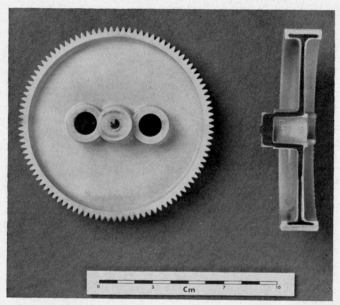

FIG. 4.25. Sandwich moulded gear wheel with p.t.f.e.-filled white nylon skin containing glass-filled black nylon core. (Courtesy I.C.I. Plastics Division.)

4.4.10   Special techniques in injection moulding

So far we have concentrated, with only the occasional deviation to illus-trate a specific point, on the moulding of normal thermoplastics materials. Injection moulding has been used since the 1930s as a means of processing rubbers and more recently other elastomeric compounds. It has also been used with increasing success for processing thermosetting polymers and expanded or foamed systems.

1. *Injection moulding of thermosetting polymers*

The processing of thermosetting materials is carried out on machines sub-stantially similar to those currently used for thermoplastics, except that a specially designed screw and barrel assembly are needed. On many modern thermoplastics machines the screw/barrel unit can easily be removed and a straightforward substitution is then all that is required.

What is needed in a thermosetting moulding polymer is a material with a plastic range extending as long as possible at the barrel temperature, followed by a fast cure at the temperature of the mould. This means that conventional moulding materials generally need some reformulation. Since they are usually more viscous than thermoplastics at the moulding temperatures, it is often necessary to add a small quantity of external lubricant such as zinc stearate so as to reduce the heat produced by the screw. Indeed it is often necessary to provide barrel cooling as well as heating in order to control the stock temperature.

Overheating can also be minimised by screw design. The material in the screw is in three forms; at the feed end and for two or three flights, it is loose powder. It then starts to fuse and becomes sintered. Finally, at the last flight it should be fully fused. Accordingly, the material feed should be fast to reduce the number of revolutions required for plasticisation. The flights at the feed end should be quite deep and the screw run slowly in order to reduce shear rate.

In general, mould design is similar to that used for thermoplastics.

2. Injection moulding of expandable plastics

The range of applications is very wide indeed but in general expanded materials are used for three reasons:

    (i) anti-sink applications—polystyrene heels, etc.;
    (ii) density reductions—reduction in weight of large mouldings, structural foams, upholstery, etc.;
    (iii) cellular mouldings—floats, cellular shoe soles, etc.

(a) *Moulding expanded thermoplastics polymers.* The type of machine used is usually of little significance in anti-sink applications. However, in the other two applications, for which considerably greater amounts of blowing agent are used, it is necessary to provide a cut-off nozzle in order to prevent premature injection into the mould. Unlike the conventional moulding technique, the expandable material is injected at a low pressure and complete filling of the cavity is accomplished by the expansion effect of the blowing agent. The effect of the cool mould surface on the hot expanding polymer causes a relatively unexpanded skin to be formed. Rapid injection is essential for good surface appearances and to ensure that expansion takes place evenly throughout the thickness of the moulding. The cell structure is dependent on a number of inter-related variables such as: type of material, melt viscosity, cylinder temperature, injection speed, type and position of gate, type of cavity, mould temperature and cooling time. The reproducibility of the moulding density and cell structures will depend on how closely these variables can be controlled and on the efficiency of the cut-off nozzle.

In view of the low injection and locking pressures needed, moulds can be made from a variety of different materials including steel, aluminium, zinc–aluminium alloys and, for prototype work, filled epoxy or polyester resins.

Since cooling times tend to be longer than in conventional moulding, it is common to combine one injection unit with several clamping units.

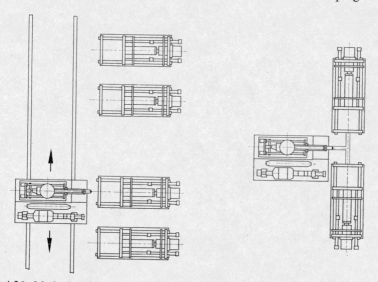

Fig. 4.26. Methods of increasing output of expanded thermoplastics polymers by combining several clamping units with one injection system. (Courtesy Siemag GmbH, Hilchenbach-Dahlbruch.)

These may either be arranged in a line (Fig. 4.26) or incorporated in a multi-station rotary press (Fig. 4.27).

(*b*) *Moulding expanded thermosetting polymers.* We shall concern ourselves here with the two principal types of foamed or expanded materials, flexible and rigid, although there is a variety of structures of intermediate character, including the products of such techniques as the sandwich moulding process, which we have already described (cf. Section 4.4.9).

As it happens much of this 'spectrum' of properties is covered by the polyurethanes (see Table 4.4) and we shall now consider the methods used processing them. Two processes are mainly used. The prepolymer system,

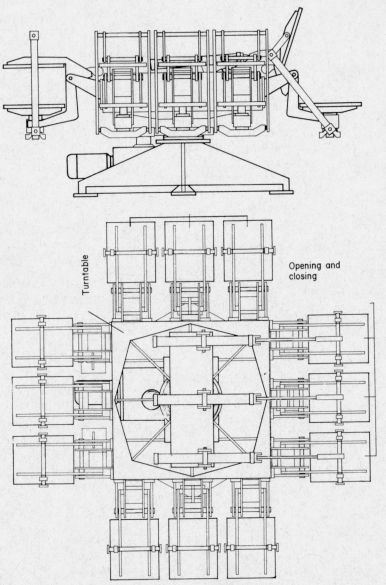

Turntable

Opening and closing

FIG. 4.27. A multi-station rotary press for moulding expanded thermoplastics. (Courtesy Siemag GmbH, Hilchenbach-Dahlbruch.)

TABLE 4.4

Type of foam	Density (g litre^{-1})	Chemical composition	Characteristics and typical applications
Soft foams	16–45	Reaction products of long chain glycols and toluene di-iso-cyanate	Hot hardening foams are open-cell materials with fine to coarse cell structure. Cold hardened foams are highly elastic with good skin formation and irregular cell structure. The main areas of application are in furniture, automotive, textile, and other household uses.
Semi-flexible foams	85–100	Based on polyols tipped with reactive ethylene or propylene oxide	Elastic but with relatively slow recovery. Extensively used in the automotive industry for arm rests, neck supports, crashpads, etc.
Integral skin	120–130	As above	Boundary layer is compact with a sharp but continuous transition to cellular core. Uses as above also in the shoe industry for shoe soles and foot supports.
Rigid foams	130–160	Products of short-chain high functionality polyols with methylene diphenylene di-iso-cyanate	Good mechanical properties and low thermal conductivity. Used in refrigerators, for structural components and furniture. This type of material can also be produced as a sandwich structure with differing skin and core properties.

which is relatively little used, where a prepolymer consisting of polyol and isocyanates in excess is mixed with a catalytic mixture of emulsifier, catalysts and blowing agents. This method is, among other things, relatively expensive. More commonly used is the one short method in which the polyol isocyanate, catalyst and blowing agent are reacted simultaneously.

Although most flexible forms are produced in slabstock form and subsequently cut to shape, an increasing proportion of both flexible and rigid foams is being moulded into shape using the one shot method. The various components are individually metered to the mixing head to allow the end

FIG. 4.28. Automatic machine for flexible foam production. (Courtesy Ed. Brand Ltd., Camberley.)

properties of the material to be determined with a high degree of precision. A machine which is capable of using up to seven components with an output of up to 350 kg min^{-1} for flexible foam production is shown in Fig. 4.28.

3. *Injection moulding of elastomers*

There are essentially two kinds of machine available for moulding natural and synthetic rubbers. These are (a) ram or piston machines, and (b) screw operated machines. As with other types of polymer the latter are more widely used now, and the main features are very similar to those already discussed.

The moulding cycle itself consists of four stages: (i) plasticising; (ii) injection; (iii) vulcanisation or curing; and (iv) ejection. Modern machines are equipped with a two-pressure control system that allows the variable injection pressure to be reduced to a holding pressure when the mould cavity is full, thus minimising flash and strain in the mouldings.

Difficulty is sometimes encountered in feeding granular stock, and provision is usually made—as with rubber extrusion—for strip feedstock to be handled as an alternative.

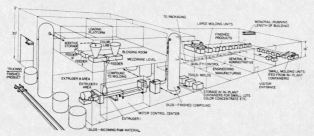

Fig. 4.29. Highly complex moulding plant with in-plant compounding facilities. (Courtesy Werner & Pfleiderer Corp.)

4.5 FUTURE TRENDS IN MOULDING

We have seen that the use of the screw has made possible a whole series of developments in the moulding field. Its application has increased the scope and versatility of the compression and transfer moulding processes, and made significant improvements in the moulding of rubbers. It is also now almost universally used in injection moulding machines.

Two main areas remain for exploitation; these are the problems of size and speed. Some of the methods used for making larger mouldings without the accompanying astronomical increases in machine costs—as well as the need for larger buildings and services to house the larger machines—have been described already. These developments are likely to continue, and we find that the benefits of cost reduction, together with increased versatility, which result from feeding two or more extrusion/injection units into a single shaping device, are becoming widely accepted in the extrusion industry also.

The problem of speed is more complex because in addition to the matter of producing—and distributing—more articles in unit time, there is the question of human attitudes.

Production rates can be increased by the wider use of automation, but—almost more important—automation can do more reliably, and in the long run more cheaply, the type of work that people are less and less prepared to undertake. Many of the tasks in industry are dull, routine and carried out in unpleasant environmental conditions. This has been recognised by the steel, automobile and petroleum industries and is beginning to find a place in injection moulding. Basically, the process breaks down into the following distinct, but closely integrated, stages: (a) raw materials handling; (b) moulding; (c) product handling; (d) storage and despatch. The thread which links all these stages together and makes full automation possible is efficient plant layout.† This in turn makes for more enjoyable working conditions and jobs with greater interest.

FURTHER READING

Elements of Injection Moulding of Thermoplastics: a Teaching Programme, Maclaren, London, 1968.

Miles, D. C. and Briston, J. H. (1965). *Polymer Technology*, Temple Press, London.

Munns, M. G. (1964). *Plastics Moulding Plant*, Vol. 2: *Injection Moulding Equipment*, Iliffe, London.

Ogorkiewicz, R. M. (ed.) (1969). *Thermoplastics*, Iliffe, London.

Processing Handbook, Bill Brothers Corp., New York, 1969.

Spritzgussteile aus Kunststoff, Krauss-Maffei, München.

† An example of one such arrangement is shown in Fig. 4.29.

CHAPTER 5

PRIMARY PROCESSING METHODS: EXTRUSION AND CALENDERING

5.1 INTRODUCTION

The ultimate purpose of the extrusion process is the manufacture of products of good dimensional uniformity and quality at economic rates of output. In this the role of the extruder is to convert the feed stock—which may be presented to it in the form of powders, granules, strip, or even

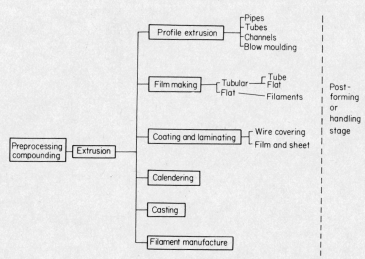

FIG. 5.1. Extrusion and extruder based processes.

melt—into a supply of homogenised melt at the correct temperature. We see then that the extruder in its capacity of mixer and melter is a vital part of a production line which comprises a number of interlinked and interdependent stages. Some of these are summarised in Fig. 5.1, and even

117

though the variety is considerable we can divide the majority of extrusion-based processes into the following components or stages:

(a) the extruder;
(b) the die—whose purpose is to shape the molten polymer;
(c) the forming stage—which sets the material to the shape created by the die or carries out a further modification to the material;
(d) the post-forming or handling stage—in which the product is collected and finished by such extra operations as trimming, cutting, reeling, etc.;
(e) secondary processing.

5.2 THE EXTRUDER

Extrusion as a method of producing continuous lengths of material is an ancient and familiar technique. It is also practised with a wide variety of materials in a number of different industries. Lead pipes, ceramic tubes and macaroni are all extruded and anyone who has squeezed toothpaste from a tube or operated a domestic mincing machine has performed two of the best known extrusion techniques: namely ram, or intermittent extrusion; and screw, or continuous extrusion.

The development of the extruder as a means of processing polymeric materials began in the years 1845–1850 when extrusion processes were used for manufacturing insulated wires and cables by coating copper conductors with gutta-percha.

These early machines were ram operated and although the technique is still used today for handling difficult materials or when exceptionally high pressures are needed, they suffered—like the toothpaste tube—from the disadvantage that the process was discontinuous. The use of several rams working in a 'cascade' sequence did to some extent alleviate this, but at the cost of more complex and expensive machinery.

The screw principle was thus developed as a means of achieving continuous processing, and the first recorded patent for an extrusion machine employing an Archimedean screw was taken out in 1879 in Britain by Gray. In the United States at about the same time, a screw-operated machine was introduced by Royle. Before the turn of the century such firms as Shaw and Iddon in this country, and Troester in Germany, had constructed and were selling screw extruders.

The early extrusion machines, like their counterparts in injection moulding, were either ram operated or were substantially similar to that patented by Gray in 1879. Being designed for gutta-percha and rubber they were required to work with hot strip feedstock, using steam for heating the barrels.

Although on the whole these machines were adequate for handling the polymeric materials available at that time, the introduction of the new synthetic polymers in the years 1920–1930 began to show up the inadequacies of these early machines. In particular, the screws then used were too short to allow for adequate heating. One of the first screw extruders which appears to have been specially designed for processing thermoplastics was produced by Troester in Germany in 1935. A few years later in 1937–1938, Francis Shaw produced a comparable machine in this country. The newer machines possessed longer screws and were heated either directly by electric heaters, or used oil as a transfer medium. Variable speed drives and automatic temperature control were also introduced at about this time, and machines produced in 1938 by Troester bear a marked resemblance to those used in the present day.

5.2.1   The single screw extruder

The one great advantage of single screw extruders over all other types is the mechanical simplicity of the single screw design and hence the relative cheapness of its construction. For this reason at least the single screw machine is still the most widely used today, and the essential features of the equipment are shown in Fig. 5.2. They consist of a specially profiled screw (1) which functions as a pump to convey the polymer—which is normally handled either as granules or in powder form—from the hopper (2) into the feed section (3) and through the heated barrel (4). During this operation the heat produced by the hot barrel walls, and the mechanical work put in by the screw, cause the polymer to melt and allow it to be forced through the extrusion die, which gives it the desired shape.

In order to provide the necessary homogeneous melt it is ideally desirable that the extrusion system should be completely stable and working in equilibrium. In practice, however, this is almost never achieved and there are variables which affect the output all along the route from feedstock to die. If we study the extrusion process more closely it will be seen that there are essentially two classes of variables which affect the quality (and quantity) of the output. These are:

(a) the variables of design, or how the extruder is made;
(b) the dynamic or operating variables which control how the extruder is run.

The effect of these two sets of variables on the efficiency of the extrusion process is complex and, as a result, a distressingly large number of extruders are run well below their maximum speeds in order to ensure trouble-free production. It is also obvious that where extruders are running well below their capacity limits, the use of the most efficient screw design becomes less important. In many operations, including blow moulding and film extrusion, the chief limitation to output lies in the time required for the formed material to cool so that it may be handled without risk of damage. Where higher speeds can be used, however, fluctuations in output do occur and both the choice of polymer and correct design of equipment are vital factors.

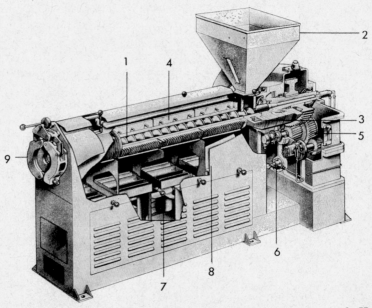

FIG. 5.2. Sectional view of a single screw plastics extruder. 1. Screw. 2. Hopper. 3. Feed section. 4. Barrel heaters. 5. Gear box. 6. Lubrication system. 7. Air blowers to control barrel heating/cooling temperatures. 8. Double walled hood for balanced air flow. 9. Die clamp assembly. (Courtesy Francis Shaw & Co. Ltd., Manchester.)

(a) *Intrinsic or design variables*

These refer to parameters whose values are selected because it is known or believed that they will ensure the most efficient operation of the system. Once fixed they cannot generally be altered without interrupting the process. They cover the geometry of the component parts which make up the extrusion system and include such important factors as the diameter and length of the screw, also the pitch and depth of the screw flights. Other variables in this category are the design of the feed hopper, die and barrel, as well as type, shape and physical state of the polymer.

The mechanical and thermal characteristics differ widely over the range of thermoplastics materials which are converted by extrusion. If the screw

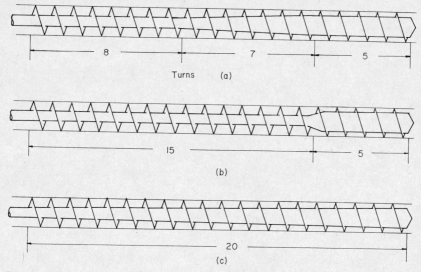

FIG. 5.3. Basic types of extrusion screw. (a) 3 Stage screw. (b) 'Nylon' screw. (c) 'PVC' screw.

is to fulfil adequately its task of providing compression, mixing and shear, as well as conveying the feedstock, it is clear that the most effective design of screw for one particular material will not necessarily be suitable for another. Ideally, of course, a separate screw should be designed for and used with each material, but since this is clearly impossible—not least on the grounds of expense—a compromise is generally made so that the screw is designed for a particular die/material combination. Figure 5.3(a) shows the profile of a typical screw which is divided into three zones: the feed zone; the compression or transition zone; and finally the metering or melt zone.

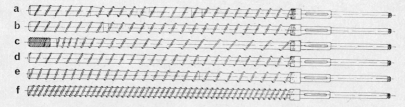

FIG. 5.4. A selection of extruder screws designed for use with the following materials: (a) Nylon (two stage screw with vent). (b) Nylon granules. (c) Reprocessing polyethylene film. (d) Polyethylene. (e) Soft PVC powder. (f) Cellulose acetate. (Courtesy Luigi Bandera, Italy.)

Figures 5.3(b) and 5.3(c) show two other screw designs which are intended to cater for specific material characteristics. Figure 5.3(b), of the so-called 'nylon' screw, is intended for use with materials with a relatively sharp melting range and low melt viscosity. On the other hand, Fig. 5.3(c) shows the 'PVC' screw designed for more thermally sensitive materials where, although adequate mixing is required, it is important not to develop too much mechanical work at an early stage during the material's journey down the barrel. Figure 5.4 shows a selection of extrusion screws intended for use with a variety of materials.

(b)　Dynamic variables

By these are meant those variables which can be altered at any time during the extrusion process. They fall into two categories:

- (i) independent variables which are controllable, i.e. melt temperature, screw and barrel cooling, and screw speed;
- (ii) dependent variables which result from the values of (i), such as melt zone position, homogeneity, output and pressure.

As might be expected screw design has a profound effect on the melt temperature. In its passage along the extruder barrel heat is gained by the polymer in two ways: internally, by friction forces caused by the mixing and compression of the screw; and externally from the barrel heaters. In the case of extrusion coating where high temperatures are required, barrel heating is almost always needed, but in film manufacture and profile extrusion, where it is necessary to keep the melt temperature as low as is practicable, screw and die design are carefully selected so that the frictional heat is just sufficient to run the system autothermally. A small amount of heat is supplied by means of barrel heaters to control the process.

Autothermal extrusion does give melt uniformity but is inflexible and can produce melt temperatures that are higher than necessary and so may require the use of barrel cooling to restore the balance. Table 5.1 shows the rise in melt temperature caused by increased shear generated by the screw

TABLE 5.1

Rise in melt temperature resulting from increased shear

Output (kg/hr)	Melt temperature (°C)
10·4	190
18·2	229
34·1	240
52·2	241
65·3	242

as autothermal conditions are approached under one particular set of extrusion conditions.

The method of supplying heat to the melt is important and, in particular, the sensitivity and accuracy of the control system. It has, for example, been shown that in some commercially available extrusion lines, temperature fluctuations as great as $\pm 15°C$ occur in the melt during extrusion. The use of induction heaters on both barrel and die is increasing owing to the fact that heating takes place directly in the bulk of the polymer rather than, as is the case with band heaters, by conduction from the heater band through the barrel wall and thence into the material. This means, therefore, that more accurate temperature corrections can be made. The effect of the temperature variation is to alter the pressure in the system, and this in turn will affect the power required to turn the screw. Modern extruders are fitted with a variety of drive systems from single or two-speed motors driving through a variable speed gearbox, to infinitely variable A.C. motors. Both A.C. and D.C. motors are also used. For applications where constancy of output is important A.C. commutator motors are generally used, although, since this type of motor is rather sensitive to changes in load, it is possible for an unstable situation to build up which encourages rather than reduces the tendency to surge. Thus it is clearly necessary to have good control of the temperature, both in the die and barrel, and also—through the motor—of the screw speed. The use of screw cooling should be mentioned here: this gives in some cases better mixing and also reduces the tendency towards surging.

5.2.2 Advances in single screw design

The power dissipated in a screw extruder varies as $N^{1+1/\nu}$ where N denotes the screw speed and ν is the 'pseudo-plasticity index' or, more briefly, the 'plasticity index'. This quantity ν indicates the degree of deviation from Newtonian behaviour.

It is interesting to note that, although output is proportional to screw speed, as is the case with Newtonian liquids, in the case of polymer melts the power dissipated in the screw channel decreases dramatically as the value of ν increases (Fig. 5.5).

In the design stage at least the screw speed is a dependent variable. Before the gearbox is fitted to an extruder, the screw speed can always be increased to obtain a desired output. The problem therefore is to design the extruder for the desired mixing efficiency for the minimum power consumption for a given output.

As screw speed increases, however, in an attempt to raise output, the residence time of the polymer in the barrel becomes less, and although the melt temperature may rise (Table 5.1) the amount of time available for mixing is also reduced.

There are two things which can be done to maintain adequate mixing at increased outputs: (a) increase the channel length, i.e. the path followed by the polymer round the screw from hopper to die; (b) increase the degree of mixing by placing a series of obstacles in the path of the melt causing it to change direction.

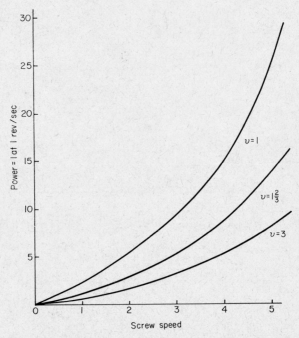

FIG. 5.5. Effect of pseudoplasticity on power dissipated in screw channel. (After B. S. Glyde and W. A. Holmes-Walker.)

Theoretical equations derived by a number of workers show clearly that the important variables in an extruder are as follows:

(1) the surface velocity V of the barrel with respect to the screw in the helical (Z) direction (see Fig. 5.6);
(2) the width of the screw channel w;
(3) the depth of the screw channel h;
(4) the helical length of the screw Z.

The obvious way of increasing the helical length Z is simply to increase the length of the screw while keeping the other dimensions constant. This is relatively simple and the L/D ratios of extruder screws have increased over the last ten years from 20:1 to approaching 30:1. A much greater increase brings complications; in the first place there is a tendency for the screw to whip causing wear in the barrel. Secondly, and this is particularly so with larger extruders, the thrust bearings required to support a cantilevered screw some 8 m long and 200 mm in diameter are massive and expensive.

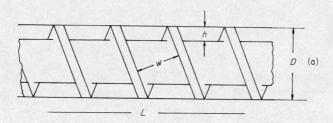

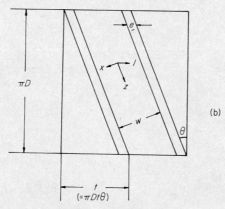

FIG. 5.6. Screw characterisation. (a) Screw dimensions. (b) Geometry of 'unwrapped' screw (two complete turns of flight).

An alternative is to lengthen Z by increasing the diameter of the screw (Fig. 5.7). If we change the L/D ratio but ensure that:

$$L_2/L_1 = D_1/D_2 = \sin \theta_2/\sin \theta_1 \tag{5.1}$$

then the performance will not be altered as long as the screw speed is adjusted according to the equation:

$$N_2/N_1 = D_1 \cos \theta/D_2 \cos \theta_2 \tag{5.2}$$

There will be a change in power requirement but the differences should be small over a wide range of L/D ratios. As an example a melt extruder with

$L = 20$, $D = 1$ can theoretically be replaced by one with $L = 10$, $D = 2$, the L/D ratio having been reduced from 20 to 5. It should be remembered that the actual load on the screw will vary as the square of the diameter so that low L/D ratios will require larger thrust bearings. A great advantage of lower L/D ratios, however, is the provision of more space inside the screw for purposes such as cooling. Here the possibility of zonal temperature control within the screw clearly exists.

An alternative method of increasing the length of the screw channel, and also incidentally of increasing the degree of mixing, was developed by Maillefer (Fig. 5.8). This screw consists basically of two channels on a common shaft, the channels being connected by a continuous passage over the bands separating them. The channel for the granules extends from the hopper and after a few turns the second or melt channel starts;

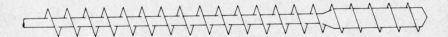

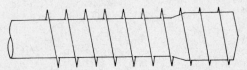

FIG. 5.7. Screws having very nearly the same extrusion characteristics.

this second channel takes over the molten material which has collected on the leading edge of the main flight by transferring it across the flight. By the time plastification is completed the granule channel ends and the melt channel takes over the whole channel width. A feature of this system is that regardless of the dimensions of the feed channel, intensive shearing and mixing starts as soon as the resin melts.

We have seen that the Maillefer screw contains both an increased path length as well as a means of increasing the degree of mixing by material transfer from one channel to another. Many devices have been tried which give rise to increased mixing, but one of the simplest and most effective is illustrated in Fig. 5.9.

The rows of pins placed at intervals round the screw serve to redirect the centre core of relatively unmixed material in the screw flight, so producing a homogeneous and well-dispersed extrudate. This type of system is particularly useful with thermally sensitive materials such as PVC.

Some materials give off gases or volatile components during extrusion. Also there is a tendency for powdered materials to entrain air into the extruder barrel. All these phenomena can have a detrimental effect on both

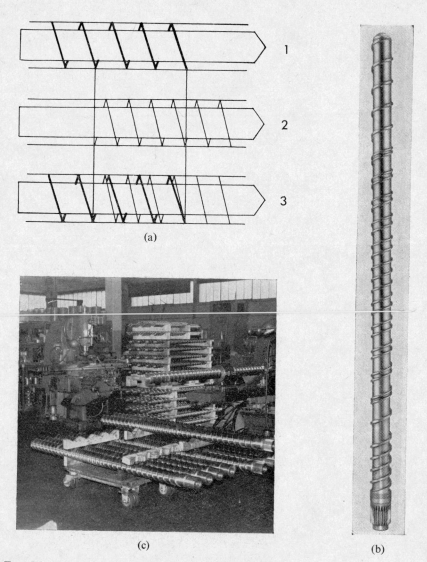

FIG. 5.8. The two channel Maillefer screw. (a) 1—entrance screw; 2—exit screw; 3 (1 + 2)—Maillefer BM screw. (b) Maillefer BM screw. (c) A selection of Maillefer screws. (Courtesy Maillefer S.A., Lausanne, Switzerland.)

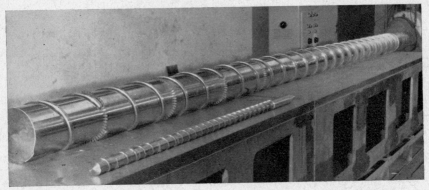

F<small>IG</small>. 5.9. Two examples of the type MS.65 extruder mixing screw showing arrangement of the mixing pins. (Courtesy Francis Shaw Ltd., Manchester.)

the output and quality of the extrudate, causing it to bubble and become porous. Air entrapment can, to some extent, be cured by using a vacuum hopper, and the evolution of water vapour can be minimised by carefully drying the polymer before extrusion.

An alternative method is to use a screw with a decompression zone (Fig. 5.4(a)). Under normal operating conditions—provided the back pressure is not too great—the flights in the decompression zone will not be completely filled with melt. It is therefore convenient to position a vent hole at that point in the barrel to allow the gases to escape. Venting is often assisted by the application of a vacuum.

A more sophisticated venting system, the Egan by-pass method, is illustrated in Fig. 5.10. The reverse flight section just before the vent opening acts as a seal. This eliminates flooding in the vent as the material has to pass through the valved by-pass, re-entering the barrel forward of the vent opening. The advantages of this system include better compounding and dispersion, wider range of materials and outputs.

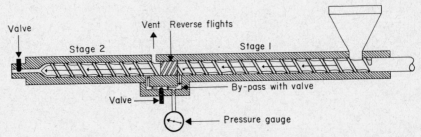

F<small>IG</small>. 5.10. The Egan extruder by-pass venting system. (Courtesy Egan Machinery Co., Somerville, New Jersey.)

Extruder output

We saw in Chapter 4 that the plasticising capacity of an injection moulding machine depends on the screw and barrel assembly with which it is fitted and the enthalpy of the material being processed (Table 4.1). Figure 4.23

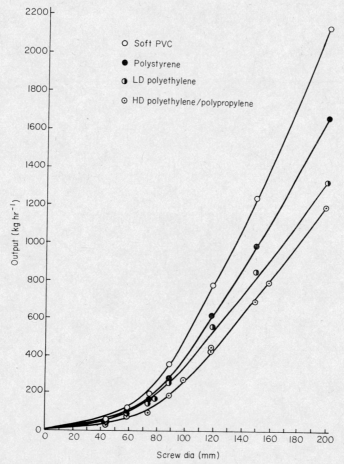

FIG. 5.11. Typical maximum outputs for extruders of different sizes with a range of polymers.

also shows the relationship between shot weight, machine cost and mould locking force.

Single screw extruders are normally described in terms of their screw daimeters and the L/D ratio of the screw; however, the level of output one can expect from a given extruder will depend—as in injection moulding—on both machine and material characteristics. As a rough generalisation, we can say that for a given extrusion system the maximum outputs obtainable at a particular range of screw speeds decrease in the following order: soft PVC; polystyrene; LD polyethylene; polypropylene and HD

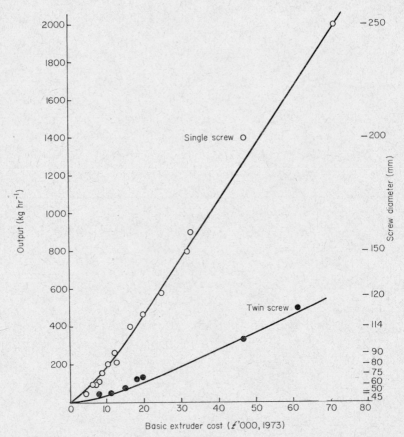

FIG. 5.12. Comparison between extruder size, cost and outputs for single and twin-screw extruders.

polyethylene. The differences become more marked as the screw diameter increases. Figure 5.11 shows performance levels for a series of extruders with L/D ratios of about 25:1 for four different polymers.

It can be seen that the spread is quite considerable. Unless otherwise specified, manufacturers usually quote maximum output values for LD polyethylene.

The use of multiple extruders is increasing for several reasons. Firstly, it is able to provide duplex structures, such as coaxial fibres or tubes and multi-component sheets; secondly, as can be seen from Fig. 5.12 extruder costs increase very steeply indeed at the higher range of outputs—particularly in the case of twin-screw machines.

It is thus often better to achieve a high output by feeding several smaller extruders into one die. In this way the factory set-up remains flexible and the extruders can be redeployed relatively easily to cater for a varied production programme.

5.2.3 Twin-screw extruders

The importance of the double screw machine has undergone certain fluctuations in and out of vogue from time to time, but recently there has been a pronounced increase in interest, to the extent that now rigid PVC compounds are processed almost exclusively by twin-screw extruders. Their use generally relies on the fact that they can process raw materials in powder form, and also on the ability of the double screw system to act as a positive action pump exerting considerable pressures.

The degree and also the pattern of mixing in twin screw machines is different from those obtained with single screws. Figure 5.13 shows photographs of cross-sections of extruded polyethylene pipe obtained with both types of system. A mixture of black and white feedstock was used in both experiments, and it is interesting to note that in addition to the more complex pattern obtained with the twin screws, an inversion of the colour distribution has also occurred.

As can be imagined, the variety of twin-screw systems is considerable: there are extruders in which the screws both rotate in the same direction and others with contra-rotating screws. An interesting extension of the conventional twin-screw concept is the use of tapered screws. The main advantage is said to lie in the fact that the surface area of the screws is greatest in the feed zone where most heat is required to plasticise the material, reducing progressively to the extrusion zone where less heat is required.

A twin-screw extruder of basically novel design is that produced by Werner and Pfleiderer using flighted bushings which are actually short lengths of screws and kneading discs. Together these add compression and kneading action to the work of the conventional screw, and the interchangeability of the bushings and discs makes possible a fine control of product quality. A similar concept is used in the GKN Windsor TS.250 twin-screw machine (Fig. 5.14), which uses interchangeable mixing, transition and metering sections.

Without taxing our imagination too greatly we can extend the concept of twin-screw extruders to include the use of two screws, each in a separate

Fig. 5.13 (*left*). Section through pipe produced with single-screw extruder using black and white feed stock. (*right*) Section through pipe produced with twin-screw extruder using black and white feed stock. (Courtesy Rubber & Plastics Research Association of Great Britain, Shawbury.)

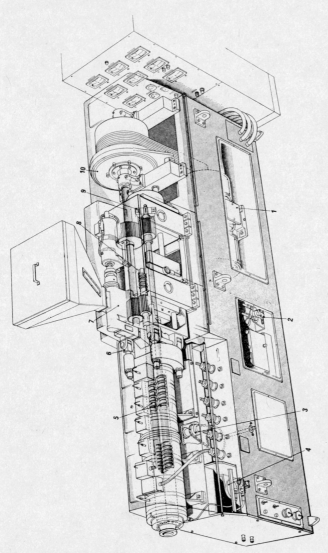

FIG. 5.14. GKN Windsor TS.250 twin-screw extruder. 1. Main motor. 2. Oil circulation pump. 3. Manually operated withdrawal mechanism for trolley mounted barrel. 4. Independent oil cooling for front barrel zone. 5. Twin screws with interchangeable sections. 6. Quick release barrel clamp. 7. Controlled material feed. 8. Multiple element screw thrust bearing. 9. Large diameter standard screw thrust bearing. 10. Epicyclic primary gearbox. (Courtesy GKN Windsor, Surrey.)

barrel—either mounted on the same frame or separately—which feed either in series or in parallel into the same die. An example of the former concept is the machine illustrated in Fig. 5.15 which contains two screws, one of which is a short subsidiary screw whose main purpose is to assist in the feeding of difficult materials. Figure 5.16 shows an example of the parallel technique which can also be used to process two different materials simultaneously.

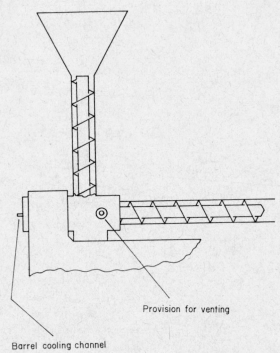

Provision for venting

Barrel cooling channel

Fig. 5.15. An example of one type of twin-screw extruder.

To sum up, each system will have its own champions and the arguments for and against are not always clear-cut. Twin-screw machines are often recommended for mixing and compounding and for their ability to handle materials which have difficult feed characteristics such as unplasticised PVC in pre-mix powder form. Also, because of the positive pumping action exerted by these extruders, they are often preferred for the extrusion of heavy cross-sections particularly in difficult materials. However, as we have seen, their disadvantages are increased costs due to their more complicated construction and the difficulty in heating materials in comparison with single-screw systems.

5.2.4   Unorthodox extrusion systems

The extrusion systems we have considered so far are those which are most widely used for plastics processing. However, it has already been stated that in order to produce the maximum theoretical efficiency the extruder screw should be designed specially for the process and material with which it is to be used.

It is hardly surprising, therefore, that special machines designed for particular purposes should appear from time to time. Some are sensible

FIG. 5.16. Dual extruder cross-head and vertical splice box used in vertical continuous vulcanisation line. (Courtesy Davy Plastics Machinery Ltd., Poole.)

and achieve lasting success, while others can really be classed as 'gimmicks' and quickly fall out of favour. A browse through the patent specifications of the last forty years or so can be quite rewarding!

Generally, we can divide these unorthodox machines in to a number of separate categories, and they are summarised briefly in Table 5.2.

5.3   THE EXTRUSION DIE SYSTEM

5.3.1   Die design

So far we have considered the basic extruder and its variations in terms of its functions as a mixer and pump. However, as has already been pointed

TABLE 5.2

Characteristics and applications of a number of unorthodox extrusion systems

Extrusion system	Description	Applications
Screw systems	Coaxial screws: outer screw rotates round inner stationary screw Plunger assisted feed to screw	Used to produce higher outputs at lower melt temperatures Increases screw efficiency by completely filling the feed section
Screwless systems	Most are based on the so-called 'Maxwell' extruder. Pressure is developed in the melt between a fixed and moving rotor. The melt then passes through one or more holes in the 'stator'. Using the Weissenberg effect reasonably high pressures and outputs can be generated	Possible uses are for easily degradable materials and those with high viscosity melts

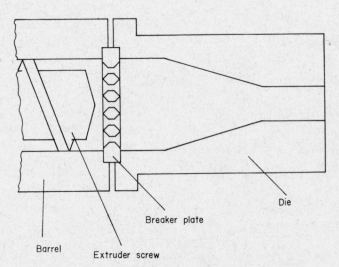

FIG. 5.17. Rod die with breaker plate.

out, the extruder is only a part of the complete fabrication system (Fig. 5.1). The melt provided by the extruder has to be shaped into the appropriate form by the die. Before entering the die the extrudate, on leaving the screw, has to pass through the breaker plate (Fig. 5.17). The function of the breaker plate is often 'backed up' by interposing wire mesh screens between it and the end of the screw. The resultant combination is thus able to: (i) filter out impurities and unplasticised polymer; (ii) provide an additional means of controlling the back pressure in the barrel; (iii) convert the rotational flow of the melt in the screw into flow that is parallel with the screw axis.

After leaving the breaker plate the melt enters the die. An extremely large range of processes and products are made by extrusion because so many different kinds of die are possible. Despite this variety most extrusion dies have a number of features in common, and it is convenient to discuss them now before dealing with the different processes.

Behaviour of the polymer melt during its passage through the die

In order to operate most successfully, all polymer conversion processes are designed so that the polymer—particularly in melt form—spends as little time as possible within the system. This is necessary for two reasons: (i) to eliminate damage through overheating; (ii) to allow production rates to be as high as possible.

The quality of the extrudate emerging from the die depends to a large extent on the geometry of the die passages; the flow should be as smooth as possible and there must be no 'stagnant' areas where polymer can be trapped and degraded.

Further, on passing through the breaker plate the polymer melt is split up into a number of individual streams which have to be recombined in the die to produce a homogeneous extrudate. This is achieved by maintaining pressure on the melt in the die by gradually reducing the cross sectional area of the die passages and by the use of adjustable restrictions in the melt path.

While travelling through the die the polymer molecules will uncoil and become oriented parallel to the flow direction. On emerging from the die orifice, however, they will no longer be restricted by the die walls, or subject to the applied pressure, and there will be a tendency for the oriented chains to resume their random arrangement. This effect—known as 'die swell'—shows itself by a contraction in the extrusion direction, accompanied by an expansion at right angles to it. There are instances where this phenomenon can be used to advantage, for example in blow-moulding, but in general it is a disadvantage and is normally compensated for or reduced by one of the following methods:

(i) increasing haul-off rate compared with extrusion rate or decreasing extrusion rate;
(ii) increasing length of die parallel;
(iii) increasing melt temperature.

These flow effects which we have just discussed need to be taken into account when designing dies for profile extrusions. For example, the effect of die swell means that the die orifice will need to be shaped as in Fig. 5.18(a) in order to produce a square cross-section, and in order to produce the cross-section shown in Fig. 5.18(b) it will be necessary to restrict the melt flow through the unshaded area either by increasing the length of the die parallel in that region, or by the use of an adjustable restrictor bar.

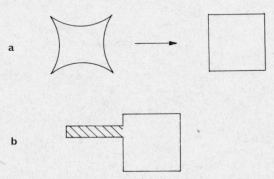

FIG. 5.18. Problems encountered in the production of solid extrudates.

Types of die

So far we have dealt with the problems of die design which are common to all extrusion processes. Although there are many of these it is possible to separate them into two general categories: (a) solid extrudates; (b) hollow extrudates.

It must be emphasised that the dies illustrated in the following paragraphs have been simplified, and that the actual dies are made so as to avoid any sharp corners, dead spots or abrupt changes in the melt flow.

(*a*) *Dies for solid extrudates.* The simplest type of die for the production of solid extrudates has been shown in Fig. 5.17, and mention has been made of the problems arising from the viscoelastic properties of the polymer melt as it passes through, and emerges from the die passages.

Another form of solid extrudate is produced when flat sheet is manufactured. Figure 5.19 shows three examples of sheet dies in common use.

In a fish tail die (Fig. 5.19(a)) the parallel is made longer at the centre to restrict the flow and so ensure that the melt emerges at a constant velocity all along the die lips. Constant thickness is obtained by the use of adjustable die lips.

The restriction to the melt in the centre of the fish tail die is sometimes insufficient to compensate for the pressure drop along the die encountered with viscous melts.

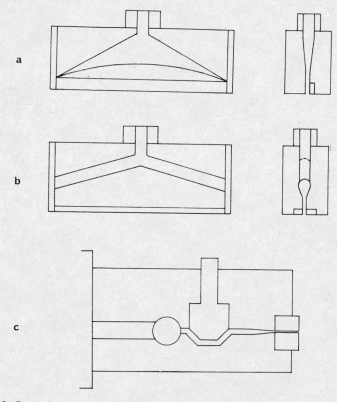

FIG. 5.19. Some designs for sheet dies. (a) Fish tail die. (b) Coat-hanger die. (c) Die incorporating adjustable flow restrictor.

One way out of this difficulty uses a shaped manifold so that the die parallel is greater near the inlet. This device, illustrated in Fig. 5.19(b) is often called a coat-hanger die.

Figure 5.19(c) illustrates a sheet die which incorporates an adjustable restrictor bar, and which may also use a coat-hanger shaped manifold.

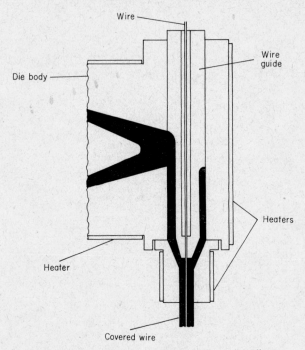

FIG. 5.20. Section through a wire-covering die.

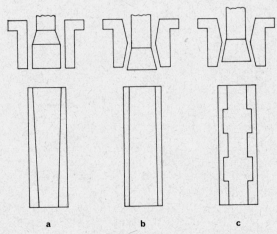

FIG. 5.21. Extrusion dies for blow-moulding. Die (a) with parallel exit produces gradually tapering parison. The valved die (b) and (c) shows the effect in (b) on the parison of continuously lowering the core, and in (c) of raising and lowering the core during extrusion of the parison.

(*b*) *Dies for hollow extrudates.* The essential features of a wire-covering die are shown in Fig. 5.20. This design is known as a cross-head die. The melt flow is interrupted by the mandrel or torpedo which carries the wire and, in order to ensure complete fusion of the melt streams, the melt temperature is kept high. Also the annular die gap is kept as small as possible and the length of the die parallel as large as possible.

The manufacture of tube is carried out either by the use of a straight-through or in-line die—similar to the rod die illustrated in Fig. 5.17 but fitted with a mandrel—or by a die similar to the wire-covering die (Fig. 5.20). In blow-moulding a specially profiled mandrel can be used which is

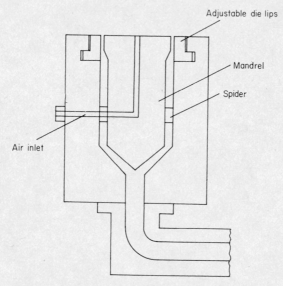

Fig. 5.22(a). Tubular film die.

moved in line with the axis of the parison to obtain the desired profile (Fig. 5.21). This axial movement alters the size of the melt passage, which in turn affects the amount of orientation in the melt. The result of this sequence shows up as a change in the amount of die-swell which takes place in the emerging parison.

A progressive reduction in the wall thickness of extruded tube if accompanied by an increase in diameter results in the production of a form of tubular film. A typical tubular film die is shown in Fig. 5.22(a). This pattern suffers from the disadvantage that the mandrel is supported by several arms. These produce weld lines (spider lines) in the extruded tube which can be sources of weakness or thickness variation. An alternative form of die is shown in Fig. 5.22(b). Spider lines are eliminated and greater melt flow uniformity at the die lips are achieved by leading the melt through a spiral channel and into a reservoir just behind the parallel section of the die lips.

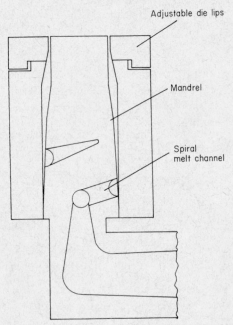

FIG. 5.22(b). Tubular film die fitted with spiral channel to ensure even distribution of melt in die gap.

5.4 EXTRUSION-BASED PROCESSES

5.4.1 Introduction

After leaving the die the molten extrudate enters the post extrusion stage whose variety extends from simple cooling and handling to the relatively complex sequence of transformation required in the production of biaxially oriented film.

Figure 5.1 shows the range of extrusion-based processes, and it is convenient to discuss them under three main headings:

A. profile extrusion;
B. film and sheet extrusion;
C. blow-moulding.

5.4.2 Profile extrusion

The variety of products which can be manufactured by this method is large, but again for convenience they can be divided into two principal categories:

(a) tubes—that is to say, hollow profiles;
(b) solid profile extrusions, in which the cross-section may have varying degrees of complexity.

This latter category also is capable of sub-division into two, namely (1) profiles, where accuracy of cross-section is relatively unimportant, i.e. draught excluders, car door seals, etc., and (2) profiles where dimensions are critical, i.e. extruded light fittings, curtain rails, etc.

The principle in all of these types of profile extrusion is, however, the same: you start from the dimensions of the final product, i.e. the sizing die, and work backwards to the extruder. The principal limiting factor in most cases is the elastic memory properties of the polymer; for example, polyethylene is generally unsuitable for those profiles where high dimensional accuracy is necessary.

(a) Tube extrusion

The general features of extruders used for this type of work have already been described in Section 5.2, but in addition some special conditions apply; for example, it may be necessary to use vented barrels in order to eliminate entrapped air. Also, the use of an efficient hopper pre-heater is common, not only to reduce the moisture content of the polymer, but also to ensure continuous production of tube of the highest surface quality at a higher output rate than is normally obtainable by cold feeding.

The barrel temperature gradient is of paramount importance in tube extrusion, but since the temperatures themselves will depend on the whole extrusion set-up, and also since opinions and techniques vary so widely, it is impossible to do more than generalise. For many applications it has been found convenient to use an inverse gradient—that is to say, with the higher temperature at the feed end. This serves to ensure closer packing of the material on the screw and produces an extrudate of greater homogeneity.

Most tube is made by straight-through extrusion, i.e. in line with the direction of extrusion with a water cooled sizing die for calibration of the external tube dimensions (Fig. 5.23(a)).

In designing a sizing die system the following factors must be taken into account:

(1) there must be provision for locating the die coaxially with the extruder die;
(2) heat transfer from the latter to the cold sizing die must be prevented;

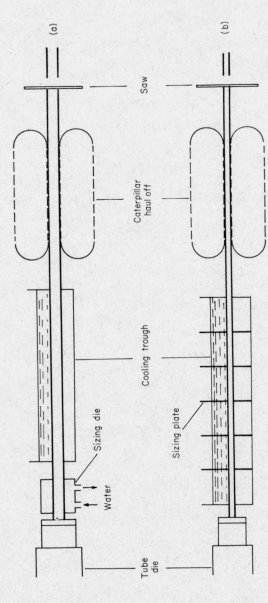

Fig. 5.23. Tube extrusion set-up. (a) Sizing die method. (b) Sizing plate method for small diameter pipes.

(3) the length of the sizing die should be related to the linear speed of production as well as to the wall thickness of the extruded section;

(4) allowance must be made for contraction of the tube which occurs on cooling, i.e. in general the internal diameter of the sizing die should be at least 2% greater than the required outside diameter of the tube.

For the manufacture of small diameter tubes sizing plates are often used instead of a sizing die, since in high speed extrusions cooling problems can occur which cause the tube to stick resulting in poor surface finish. A typical arrangement of sizing plates is shown in Fig. 5.23(b).

The post-extrusion stage in tube manufacture is extremely important although much of the information on the effect of such important parameters as the temperature of the cooling water, the relationship of die swell and the final stage of the profile, and also whether or not the tubes should be annealed during or after the manufacturing process, has been amassed largely by empirical means.

Some of the factors contributing to poor products are shown in Table 5.3.

TABLE 5.3

Factors contributing to the production of poor quality tube

Cause	Effect
Materials temperature uneven or too low	Tube external surface dull. Rough pattern on bore
Sticking in sizing die due to over-heating	Drag marks on external surface
Slight sticking in sizing die or surge in extruder output	Ripple in bore
Moisture in compound	Rough pattern on bore and/or voids in tube wall

(*b*)   *Wire and cable covering*

This process is analogous to tube extrusion on two counts. In the first place we may visualise the structure of the coated wire as being a tube whose interior is filled with metal instead of with air. In the second place both processes use a torpedo: the wire passes through the torpedo and is coated with molten polymer before it leaves the die head (Fig. 5.20). A typical arrangement of the equipment for producing polyethylene coated wire is shown in Fig. 5.24. It must be appreciated that this set-up only shows the basic features of the process and that a full scale high speed

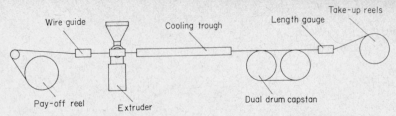

FIG. 5.24. Diagrammatic view of insulating and sheathing line.

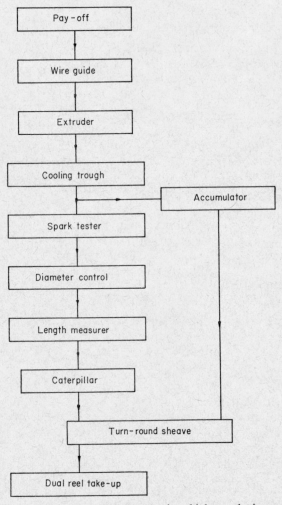

FIG. 5.25. Schematic view of the elements in a high speed wire-covering line.

coating/sheathing line is likely to contain the components shown schematically in Fig. 5.25. The complexity of the set-up is due to the need for producing an uninterrupted supply of coated wire at very high speeds (with suitable equipment coating speeds in excess of 1800 metres per minute can be achieved). In order to make this possible, dual pay-off and take-up systems are used and particular attention is paid throughout the line to the elimination of wire stretch. In cases where the economics of the process can justify it the pay-off reels can be replaced by a wire drawing machine and continuous annealer.

Fig. 5.26. Part of a high speed wire coating line showing the caterpillar and horizontal accumulator. (Courtesy Francis Shaw & Co. Ltd., Manchester.)

As speeds increase it is obviously necessary for the length of the cooling zone to be increased and this is sometimes achieved by the use of a combined capstan and multi-pass water trough. In order to facilitate changing from one take-up reel to the next an accumulator is used which incorporates a moving sheave (Fig. 5.26).

Under normal operating conditions with thin gauge wire the linear covering rate is governed by the speed of the capstan reel, although particularly in the case of thicker cables the maximum output of the extruder is the limiting factor.

Additional factors in determining the production rate are the length and temperature of the cooling bath, the type of polymer and the thickness of the polymer coating. In order to obtain a uniform coating which adheres tightly on the wire it is usually necessary to draw the melt down to compensate for the effect of die swell and to position the tip of the torpedo slightly behind the die parallel (Fig. 5.20).

In addition to wire covering in the manner already described, it is often necessary to produce insulated and sheathed cables using rubber compounds or cross-linked polyethylene, and provision has to be made for the wire to pass through a vulcanising tube. Continuous vulcanisation lines in general include the stages illustrated in Fig. 5.25 and include vertical, catenary and inclined systems.

Among the factors which will dictate the choice of the particular system are the weight of cable involved, whether or not the amount of unsupported length would cause stretching, and whether or not short or long runs are envisaged. For example, a vertical line is relatively easy to start up compared with a catenary line and short lengths can be processed relatively easily. However, as against this the installation costs are expensive due to the height of the tower.

(c) Extrusion of solid sections

Generally speaking the process arrangements are similar to those already discussed for the production of pipe except that obviously it is no longer necessary to inflate the extrudate up to the correct size.

Many of the polymers which are used for the production of solid sections pass through a fairly extended rubbery stage on cooling, and for this reason it is possible to carry out a range of post-extrusion operations; for example, polymethyl methacrylate light fittings are made by slitting the extruded tube as it emerges from the serrated die. It is then drawn over a series of shaping mandrels after which the extrudate is cooled. In addition it is possible to alter the profile of the extrudate by passing it between shaping rollers. These may also be used to impress a pattern on the product. Finally, of course, it is possible to produce extrudates with areas of different colours by using two or more extruders leading into a common die.

5.4.3 Film and sheet extrusion

The boundaries between film and sheet are difficult to draw with any precision but it seems generally to be accepted that material up to 0·25 mm thick is film and above 1 mm is sheet. The middle range (0·25–1 mm) is normally described as foil and is technically capable of production by the methods and equipment used for both film and sheet, although generally speaking it is uneconomical to use large equipment running well below capacity to produce the thinner material.

Although there is this overlap in techniques in the middle, the methods used in production are different for film and sheet. Film can be made by two systems, the tubular process and the flat film casting method. Sheet, on the other hand, is almost always made by the flat process.

FIG. 5.27. Blown film extrusion set-up. (Courtesy Bone Cravens Ltd., Sheffield.)

1. The tubular film manufacturing process

In the manufacture of tubular film the cylinder of molten polymer after leaving the die is inflated to the required diameter by air pressure. The inflated film is then cooled and collapsed between nip rolls prior to being wound up, or slit and then wound up (Fig. 5.27). With thin film manufacture, as we have already mentioned, one of the principal limitations is the removal of heat from the inflated tube. Unfortunately, air is an inefficient heat transfer medium; refrigerated air and water have been tried also, and Fig. 5.28 shows a recent development intended to secure greater production speeds. With thicker films the limitation is often the liability of the extruder to surge at higher outputs. The suitability of films for particular end uses will obviously be governed by their properties. For most applications it is sufficient to consider the following properties, although their relative importance will differ according to the end use:

(a) thickness;
(b) width;
(c) mechanical strength;
(d) surface and optical properties;
(e) orientation.

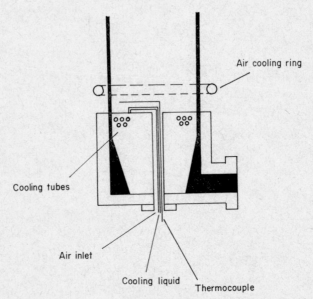

Fig. 5.28. Mandrel cooling system (introduced by Goulding Plastics). The circulation of a coolant through the die mandrel is claimed to increase outputs by up to 10%.

The effect of process variables on the properties of layflat film: (*a*)   *Thickness.* Film thickness in the first instance is controlled by die geometry and the temperature distribution over the die area. However, once an extrusion line is running satisfactorily it forms a very stable system, and in view of this the ultimate film thickness is controlled by making adjustments to the blow-up ratio and haul-off speed. If the die design is good and the die gap uniform, then uniform cooling of the bubble should produce uniform film. In practice, however, there are always small variations in the setting of the die gap as well as in the cooling devices, both of which may cause local thick or thin spots in the film. Although for many applications this type of thickness variation is tolerable, a roll containing uneven tensions will result when the film is wound up. Good rolls can be obtained

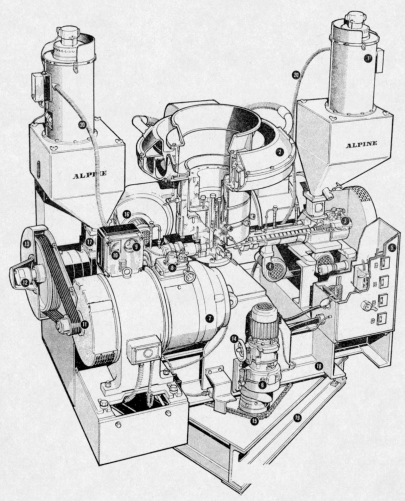

FIG. 5.29. Vertical twin-extruder blown film plant. (Courtesy Europlastics Monthly.)

1 Pneumatic hopper loader	8 Motor speed controller	15 Chain drive for rotary table
2 Air cooling ring	9 Tachometer	16 Centrifugal fan for cooling ring
3 Quill drive	10 Ammeter	17 Feed control valve
4 Barrel cooling fans	11 V-belt drive	18 Rotary table frame
5 Control cabinet	12 Screw (shear) gap adjuster	19 Base frame
6 Low geared motor for table drive	13 Safety guard for drive	20 Hopper feed tube
7 Extruder drive motor	14 Rotary table speed control	

by dispersing the high spots, and this can be achieved by rotating or oscillating either the die or cooling ring—or both—at the same time. Figure 5.29 shows part of a blown film plant which uses the oscillating die principle, and incidentally uses twin vertical extruders feeding into the same die. Both methods are successful, but it has been observed that a rotating air ring gives generally more even gauge than does a rotating die. Most tubular film dies currently in use consist of two comparatively massive pieces of metal—a central core within an annular die lip—and any adjustment to reduce the size of the thick spot on one side will cause increased thickness on the opposite side. Provided the circumference could be uniformly heated, more uniform tube could be attained by the use of a continuously deformable circular die lip. Both film thickness and layflat width are affected by variations in haul-off speed and blow-up ratio. The former is essentially a matter of engineering in that methods of speed control are well developed and you get what you are prepared to pay for.

(*b*) *Width*. Bubble width, on the other hand, is strongly temperature dependent, and it is possible to minimise the effect either by surrounding the bubble with a thermostatically controlled environment, or by ensuring that the air inside the bubble remains at a constant temperature. Alternatively, it is possible to keep the width constant by the use of a continuous monitoring system which controls the air pressure in the bubble.

(*c*) *Mechanical properties*. Although many of the experiments from which the following conclusions are drawn were carried out some years ago by B. S. Glyde in the Metal Box Company laboratories in 1959 and 1960, and by Clegg and Huck in 1960 and 1961, they are still applicable since the changes that have taken place in recent years are refinements in techniques which have improved the process without altering the basic pattern.

Although not strictly a variable in the present context, the choice of the appropriate polymer is particularly important in achieving the desired end properties and many compromises have to be made. For example, in the development of intermediate density polyethylenes the flow properties have to be adjusted to give high clarity film, and the density increased at a sacrifice of toughness to give the required non-blocking characteristics. Use of additives also can affect processing behaviour.

When considering the mechanical strength of film the important factor is the likely durability of the film in service, and this factor is almost impossible to assess in a simple way. However, it is accepted that a reasonable simulation of the necessary properties can be obtained by measuring the three quantities—tensile, impact and tear strength. The effect on properties of the conditions obtaining in the actual drawing zone between the die and the blowing ring are important, but also very complex. Therefore, they will be dealt with rather superficially. There are six important variables: melt temperature, film thickness, haul-off speed, layflat width, die gap, freeze line height.

A general summary of the effect of these variables on the mechanical properties is shown in Table 5.4. Briefly, it can be said that:

(i) For high impact strengths the blow-up ratio should be between 2 and 4 with as high a freeze line as is consistent with bubble stability. The effect of keeping the freeze line high is accentuated at higher blow-up ratios and is more marked at the higher melt temperatures, although it is possibly necessary to make a compromise with haul-off speed. Impact strength is in fact considerably increased by higher haul-off speeds, but at these higher outputs the bubble shape changes so that orientation in each direction instead of occurring simultaneously is now separated in time; the sideways one taking place later and just prior to the freeze line. An increase of 20% in die gap was found to increase the impact strength by about 80%.

(ii) For high tear strength it seems well established that it is extremely difficult, if not impossible (for a given polymer), to obtain high impact strengths combined with high tear strengths. Here, a lower blow-up ratio is preferred with high freeze line together with a larger die gap. These rather conflicting findings are also confirmed by I.C.I.'s work.

TABLE 5.4

The effect of different process variables on the mechanical properties of L.D. polyethylene

| | Mechanical property | | | | |
| | Tensile | | Tear | | |
Factor	Machine	Transverse	Machine	Transverse	Impact
Die gap	+ +	+	+	0	−
Melt temperature	0	0	−	+	+
Blow-up ratio	− − −	+ + +	1	+	+
Freeze-line height	−	+	−	+	+
Haul-off speeds	0	0	0	−	0

Note. Results are shown as a change in properties following a change from the lower to the higher level of each factor.

(d) *Surface and optical properties.* The two major types of optical irregularity in polyethylene films are caused by:

(i) surface irregularities resulting from melt flow phenomena; and
(ii) crystallisation behaviour.

During the film forming operation there will be a change in the texture of the surface of the melt because of a decrease in the depth of the extrusion defects, and an increase in their length and breadth as the melt is drawn lengthways and sideways. As long as the polymer remains molten, the overall magnitude of the defects will decrease under the influence of surface tension. Thus, as the freeze line height is increased the haze will diminish, but at the same time as this is happening the opposing effect of the slower cooling times on the size of crystallites will also operate. Generally speaking, blow-up ratio, freeze-line height, die gap, and extrusion temperature, should all be kept as high as possible. These conditions are, except for die gap, similar to those needed for high impact strengths.

(*e*) *Orientation.* Due to their method of production all blown tubular films are oriented to a greater or lesser extent depending on the processing conditions. Systems have been designed to increase the degree of orientation in order to obtain films of improved clarity, strength and heat resistance. Except for the special applications where greater strength in one direction may be needed, films are normally made with balanced properties, and this is achieved by producing film with equal orientation in both the machine and transverse directions. This does not mean, however, that the amount of draw-down and the degree of inflation of the tube should also be equal, since the amount of orientation produced by these two operations is not the same.

If the film is kept in the molten state for some time after it has been stretched, the molecules will tend to slide past one another and the stretch-induced orientation will decrease as a function of time and temperature. If, on the other hand, the temperature is kept as low as possible during and after stretching, highly oriented films can be produced which are many times stronger than the unoriented variety (see Table 1.3).

Orientation can be induced in either flat or tubular films and complete lines for the production of oriented films are now available. In general, all the important factors already discussed in film manufacture apply in the case of oriented films but with increased effect. For instance, thickness-control of the extrudate is particularly important, since the subsequent orientation stages tend to magnify variations. Control of variables in the drawing and post-orientation stages is also of especial importance since the whole time/temperature relationship, including rate of drawing, film temperature and so on, will profoundly affect the film properties. Uneven heating will affect the working of both processes but especially the tubular one since bubble instability will result. One other problem, not usually encountered with the rubbery films like polyethylene, but significant in materials such as polypropylene and polyethylene terephthalate, is the difficulty of transforming a tube into a flattened double (or slit) sheet.

2. The flat film extrusion process

In this process the curtain of melt issuing from the slit die is allowed to fall vertically into a cooling system which may consist of either (1) a water bath, or (2) a chill roll assembly (Fig. 5.30). The water quenching process has serious limitations when high production speeds are attempted since water may be carried over from the cooling bath making it impossible to dry the film sufficiently for it to undergo printing pre-treatment.

In the chill roll casting process, however, the limitation is simply the rate at which heat can be removed from the film.

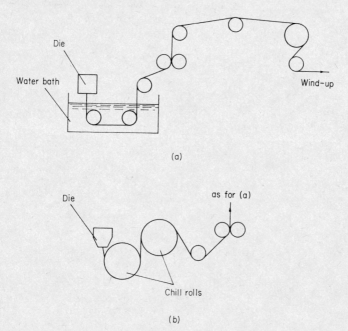

FIG. 5.30. Different cooling systems used in the manufacture of flat film. (a) Water bath. (b) Chill rolls.

The parameters which affect the properties of cast or quenched film are summarised in Table 5.5.

The most important aspect of the film making process is to ensure even contact of the hot melt on the first chill rolls. At speeds greater than about 15 metres per minute the melt tends to lift away from the roll and ride on a pocket of air. To overcome this the melt is held on to the roll by means of an air-knife system, which uses a fine jet of air to hold the hot plastic melt on to the chill roll. Care must be taken to avoid turbulence effects which will cause vibrations of the hot melt.

Polymer Conversion

TABLE 5.5

Some parameters affecting the properties of extruded polyethylene film

Variables	Level	Effect on film properties
Melt temperatures	High	Reduced haze and improved gloss, higher outputs
Cooling rate	High	
Air gap between die and quench roll/bath	Low	Reduced haze and improved gloss, low orientation, improved melt strength

Although the flat film method offers advantages in higher speeds the tubular film process is obviously capable of producing wide film and, in deciding on the method to adopt, these different and often conflicting factors must be taken into account.

Sheet extrusion. In the manufacture of sheet a horizontal extrusion system is normally used (Fig. 5.31) and, due to the larger hot mass of film, air-knives are almost invariably fitted.

The application for which the sheet is made will to some extent govern the conditions of manufacture.

For example, where the material is intended for thermoforming it is often slightly stretched during manufacture so as to prevent sagging and drooping during the thermoforming process. Generally speaking, however, balanced properties—to within 10%—are aimed at, and care is taken to ensure that the speeds of the various parts of the system are carefully matched. Nevertheless, since the hot melt has been extruded through a die system, there will inevitably be a residual memory effect which may cause the sheet to bow either upwards or downwards in both

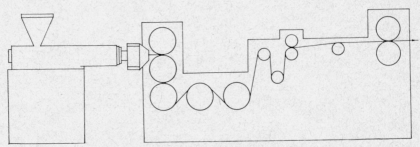

FIG. 5.31. A high speed extrusion line for the production of polystyrene sheet. Fitted with a 150 mm extruder, outputs of about 1000 kg hr^{-1} are usual for gauges in the range 0·25–1·25 mm.

the machine and transverse directions. These tendencies are controlled by adjusting the temperature of the upper and lower polishing rolls and also by controlling the temperature gradient along the rolls.

The manufacture of expanded sheet and film. Two methods are generally used: (i) moulding; (ii) extrusion. In the moulding process—which is particularly used for the manufacture of thick sheets (> 10 mm)—the moulded block is slit into sheets with hot-wire knives.

There are two variants of the extrusion process available; the first being essentially a continuous version of the moulding method, except that the foamed material is allowed to expand freely into air instead of into a closed mould. A 'log' of expanded material emerges from the die and, after cooling, is cut into sheets with hot-wire knives in the same way as the moulded blocks.

The alternative extrusion-based method produces a tube of expanding melt from the die, which is then either inflated with air or drawn over a mandrel (Fig. 5.32). This inflation stage, which is similar in principle to that used in the manufacture of layflat film (cf. Section 5.3.3), is vital to the successful production of expanded sheet and film for another reason. As the melt is under pressure when it emerges from the die lips it will continue to expand in all directions. The longitudinal expansion can be accommodated by increased haul-off speeds, but expansion in the other two directions tends to introduce corrugations into the web which can only be removed by stretching. It is for this reason that it is impracticable to make a flat sheet from a slit die. In the case of the tubular material a stretching ratio of about 2·8:1 is found to be sufficient to remove the corrugations. Although more cumbersome, since it involves the provision of a range of mandrels of different sizes, the mandrel method eliminates edge-trim waste since the tube is slit immediately on leaving the mandrel (Fig. 5.33).

Attempts have been made to produce expanded films from polymers other than polystyrene by extrusion, but problems of melt strength and the characteristics of the available expansion systems have not yet been overcome on a commercial scale.

Process variables. The pioneering work on the extrusion of expanded materials was carried out on single screw machines of about 20:1 L/D using gassed granules. This, although capable of producing material of adequate quality used an expensive form of feedstock, and two different approaches were developed. The first used a twin-screw extruder (L/D about 15:1) with gassed beads as the feedstock. This form was relatively cheap as it could be taken directly from the reactor stage. Again, although successful, it was found to be economical only at outputs up to 350 tonnes/ year.

FIG. 5.32. The production of lay-flat film from expanded polystyrene. (Courtesy Bone Cravens Ltd., Sheffield.)

For higher outputs a long single-screw extruder (L/D about 31:1) with ungassed granules proved more efficient (Fig. 5.34). With this technique the expansion system is injected through the barrel wall into the flights of the screw. A further modification developed by the NRM Corporation (Fig. 5.35) is to use two single screw extruders in tandem. The first screw carries out the operations of gelling and mixing, and the second continues the mixing sequence and cools the material.

As with all extruder-based manufacturing processes, the successful achievement of the product depends on the continuous interaction of a

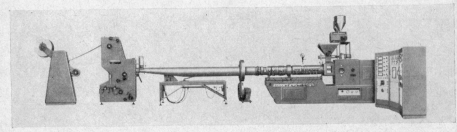

FIG. 5.33. General view of line for the production of expanded polystyrene film using mandrel method. (Courtesy Lavorazione Materie Plastiche, Turin, Italy.)

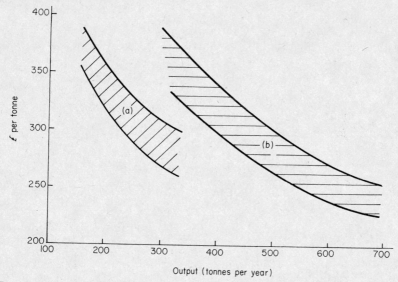

FIG. 5.34. Cost of expanded polystyrene sheet manufactured by two processes. (a) Twin-screw machine using gassed beads. (b) Single-screw long extruder using ungassed granules. (Courtesy Shell Chemical Co. Ltd.)

number of process variables (cf. Section 5.2.1). The two principal criteria for the assessment of performance are: (a) appearance; (b) maximum expansion (to ensure greatest rigidity/unit weight), and these are satisfied by ensuring the most appropriate balance between the following variables: (i) mass temperature; (ii) gas type; (iii) gas content; (iv) molecular weight; (v) number of nucleating centres.

Coating and laminating: (a) *Extrusion coating.* An extrusion coating line consists of these basic units: the unwind station; the coating

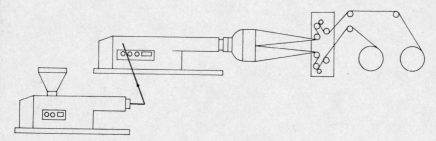

FIG. 5.35. The manufacture of expanded polystyrene film using two extruders (after NRM Corporation, Akron, U.S.A.).

and laminating unit; and the wind-up unit (Fig. 5.36). The process itself
is complex containing many variables each affecting the quality of the
coated product, and the interaction of these variables is of considerable
importance.

The objective in extrusion coating is to adhere the relatively thin polymeric coating (usually of polyethylene but occasionally of polypropylene,
and more recently of nylon and various other materials) of uniform gauge
to a substrate which may be either of metal, paper or plastic. This is
achieved by extruding a thin film of molten polymer from a slit die. This
film is drawn down to its final thickness in the air gap between the die lips
and the point of contact on the substrate. Adhesion between the substrate
and the film is promoted by pressing and quenching between the pressure

Fig. 5.36. Extrusion coating line. (Courtesy Bone Cravens Ltd., Wembley, Middx.)

and chill rolls. A variety of chemical adhesion promoters is also often
used. Due to the presence of the relatively stable substrate and the rapid
cooling which occurs, very high speeds can be obtained in the extrusion
coating process. As we discovered in the case of wire covering provision
has to be made for reel changing without interrupting the continuity of the
process and also, since breakages sometimes occur, it is necessary to provide the facility for suddenly stopping the passage of the substrate and re-
threading. These requirements, together with the need for precise speed
control at high speeds, mean that coating equipment is costly, and it is
common for complete coating lines to be supplied by the manufacturer.

The important factors to be considered in the extrusion coating process
are as follows:

(i) adhesion between plies; (ii) odour; (iii) heat seal range; (iv) gauge
variation; (v) neck-in; (vi) surface; (vii) dimensional stability.

Changes in dimensions in the finished laminate can arise from two causes:

(a) by changes in the substrate due to overstressing during passage through the machine, by uneven heating effects, and by gains or losses in moisture content;
(b) by unbalanced forces in the coating.

Both show up either in storage or after a post-extrusion treatment which tends to release the stresses. Some promise has been shown by attempts to alter the orientation in the melt by adjustment of the screw speeds and take-off speeds, to maintain the desired coating thickness. (Adjustment of both variables is usually needed.)

Coating lines, it will be appreciated, take some time to run up to full speed and to reach equilibrium. They are therefore best suited to long production runs.

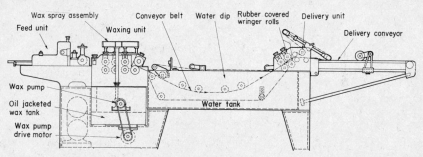

FIG. 5.37. Diagrammatic view of a wax coating line for the production of high gloss carton blanks. (Courtesy Bone-Cravens Ltd., Wembley, Middx.)

(*b*)   *Melt coating.*   In addition to extrusion coating two other coating techniques are commonly used. These are: solvent coating and melt coating. The operating costs of the two techniques are approximately the same when the cost of solvent recovery is included in the solvent coating process. These costs are 25–50% greater than extrusion coating and about one-quarter that of a calender line.

The main advantages of the melt roll coating process are versatility and flexibility and the ability to operate short runs efficiently for a wide range of end products. This means that under normal factory conditions, which often require colour changes necessitating a considerable amount of down time on solvent coating and extrusion coating processes, the total production of the melt coating process is almost double that of solution coating and from 20–25% greater than extrusion coating for coatings exceeding 0·05 mm in thickness. Figure 5.37 shows a typical melt coating production line used for the production of waxed cartons.

FIG. 5.38(a). A multilayer sheet plant based on two 90 mm extruders, one 60 mm and one 45 mm, feeding into a four layer composite sheet die. (Courtesy Reifenhauser K.G., Troisdorf, W. Germany.)

FIG. 5.38(b). View of die used in the production of multilayer sheet. (Courtesy Reifenhauser K.G., Troisdorf, W. Germany.)

(c) *Laminating.* We have just seen that structures containing a number of plies can be made by one or other of the several coating techniques. However, it is also possible to produce such structures by laminating. The appropriate collection of films is bonded together by using an adhesive, by solvent bonding or simply by the application of heat and pressure.

The various plies can be brought to the laminating station either from reels or as films which have been extruded and cooled just prior to the laminating stage. Figure 5.38 shows the production of a complex four-ply laminate in which a number of extruders feed into a single compound die. Although in this example the extruders each handle different materials it should also be borne in mind that for the reasons discussed in Section 5.5.2 it is often convenient in the film and sheet making processes for several smaller extruders to feed into the same die to achieve the same output that would be produced by a single, large, and very costly machine.

5.4.4   Extrusion 'blow moulding'

This most versatile technique is widely used for the fabrication of hollow articles. Its principal application is in the manufacture of containers, and the ranges of sizes extend from domestic fuel tanks of 2000 litres capacity down to small pharmaceutical phials of 2 ml or less.

In its simplest form blow moulding, a process borrowed from the glass industry, consists of clamping a hollow thick-walled extruded tube—or parison, as it is called—between the halves of a hollow mould and inflating the parison into the closed mould (Fig. 5.39). If the parison extrusion is intermittent, the only movement required by the mould is that of opening and closing. If, as is more usual, the extrusion is continuous, the process operates either (a) by severing and transferring the parison to the mould, where inflation and cooling takes place, or (b) by moving the mould away from the extrusion head while the parison for the next cycle is being extruded.

The varieties on the basic process are very numerous. For example, bottles can be blown horizontally, right way up, or upside down, and bottle necks can be formed in a variety of ways. We shall, therefore, restrict ourselves to considering only the significant features of the more familiar techniques.

Methods of increasing the output from blow moulding equipment have been mainly along two lines, depending to some extent on the nature of the polymer to be used, and also on the economics of production. The methods are: (i) the use of multiple parisons and a corresponding number of moulds; (ii) the use of a single output die in conjunction with a series of moulds.

(*i*) *The multiple parison system*

Although capable of excellent results this method needs careful control, possibly by the use of adjustable restrictors, to ensure that all parisons are of equal length; uneven parisons lead to variations in bottle weight

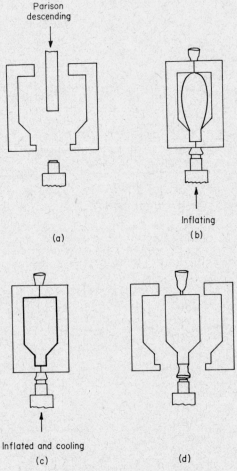

Fig. 5.39. Stages in the blowing of a plastics bottle.

and also generate more scrap in the 'ears' and 'tails'. A further improvement in output can be achieved by using two sets of moulds which operate alternately thus ensuring the maximum possible cooling time in the mould.

Figure 5.40 gives a schematic view of this device applied to a four-parison system.

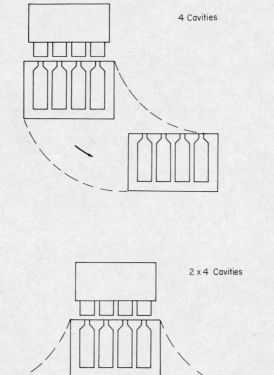

FIG. 5.40. Method of increasing output rates by using two alternating pairs of cavity assemblies. (Courtesy Kautex-Werke, Reinold Hagen.)

(*ii*)   *Single die/multiple mould system*

In this method the moulds are carried on a revolving chain or mounted on a rotating table, and moved either vertically or horizontally. Production rates are impressive: for example, 18 moulds mounted on a vertically rotating wheel can produce more than 6000 bottles per hour.

Both techniques use continuous extrusion and further improvements in output can be achieved by mounting the extruder so that it oscillates towards and away from the moulds. In the case of the multiple mould system this allows a continuously rotating mould table to be used more effectively without the complex engineering requirements for stopping and starting the relatively massive assembly of moulds as each approaches the parison.

Production of special categories of mouldings: (*a*) *Large mouldings.* A parison extruded downwards will start to extend and become thinner under its own weight as extrusion continues. This sets a limit to the size of container that can be made, since the larger the parison the more pronounced becomes the thinning and elongation. A way out of this difficulty is to extrude the parison so rapidly that thinning is minimised. This is achieved by the use of an accumulator, which consists of a reservoir (whose capacity can exceed 100 litres) into which melt is fed by the screw, and from which it is forced rapidly through the die by means of an hydraulic ram. Mouldings of up to 2000 litres capacity are achieved by this method. Several extruders can be used to provide the necessary supply of melt (Fig. 5.41), and for special applications it is possible to use two accumulators feeding into the same die.

(*b*) *Small mouldings.* Small containers, e.g. for medicinal use, of 1–5 ml capacity can be produced by a stamping technique in which the parison is 'flattened' between the two halves of a multiple cavity mould. The small bottles or phials are then punched out mechanically or trimmed by hand.

(*c*) *Miscellaneous mouldings.* A wide variety of techniques has been introduced into the basic blow moulding process in order to increase its versatility. It is only possible to discuss some of them here.

Various devices are used to obtain as precise a neck form as possible; these include drilling the necks after forming, compression forming necks, and even the use of preformed neck sections which are inserted into the mould prior to blowing.

Complex shapes are also made by blow moulding several components at once and then separating afterwards. This technique is used in the manufacture of ventilation and heating ducting for cars.

Flat containers, and containers with handles or off-centre necks, can be made by deforming the parison just before the mould closes; this is achieved by the use of moveable blowing spigots or fingers which stretch or bend the parison in such a way as to ensure the most uniform wall thickness over the whole container.

The achievement of correct material distribution in a moulding is important for two reasons: (i) to produce adequate strength and rigidity; (ii) to eliminate waste. Two factors, however, make this control difficult with the simple parisons so far described. The first is the progressive cooling and draw-down of the emerging parison, which means that not only will the parison be thicker at its lower end, but also cooler. Secondly,

FIG. 5.41. An extrusion blow-moulding machine for the production of large containers. The machine is fitted with four extruders and the accumulator is complete with automatic parison programming for wall-thickness control. (Courtesy Kautex-Werke, Reinold Hagen, W. Germany.)

the portion of the inflating parison which first touches the inside of the cooled mould will solidify, and subsequent blowing and thinning will occur elsewhere.

The most widely used technique for controlling material distribution is that of 'valving' or producing a parison in which the material in the wall is varied with time during extrusion. The method was described in Section 5.4.3 and the features of a 'valved' die are illustrated in Fig. 5.21.

Three other important developments have also occurred over the last ten to fifteen years; the use of pre-formed parisons, the introduction of 'parison-less' blow-moulding and the development of the injection blow-moulding process.

The use of pre-formed parisons. Introduced originally in the 1960s by the Marrick Engineering Company, this process uses parisons cut from a continuous length of extruded tube. These parisons are mounted on blowing spigots, heated and blown into bottles when required. Advantages are that bottle manufacture can be carried out away from the extrusion stage

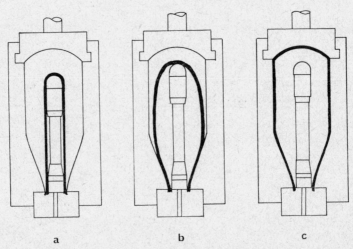

a   b   c

FIG. 5.42. The production of oriented bottles by the VPM process. The preform has been closed and heated and carried into the mould. (a) The preform is in the mould; (b) the mandrel has moved up to bring the preform into contact with the top of the mould cavity and the first part of the blow has started; (c) the blow is completed. (Courtesy Plastics and Rubber Weekly.)

(for example, as part of the bottling plant), also the extruder can be used for other production once the required length of tubing has been made.

This process has been extended recently by VPM, a German consortium, for the manufacture of bottles suitable for holding carbonated and pressurised drinks. These present two problems: mechanical strength and permeability. The VPM process starts with an extruded thick-walled tube. The ends of the tube are heated, one end is closed and the other formed into a specially shaped neck (Fig. 5.42(a)). The parison is next heated selectively so that those parts which will have the greatest stretch applied receive the most heat. Just before blowing, the heated parison is stretched by raising the mandrel to impart a high degree of molecular orientation and blowing then takes place (Figs. 5.42(b) and (c)). The process is said to be capable of producing both crown and screw top bottles at rates of up to 10 000 per hour.

Blow-moulding without a parison. This method, introduced in the summer of 1973 by Siemag, is mid-way between conventional blow-moulding and injection blow-moulding. Basically the process uses a non-stick coated mandrel which is dipped into molten plastic. This material, in the form of a sleeve, is transported to a normal blowing mould, removed from the mandrel, and blown. The process is illustrated schematically in Fig. 5.43; differential material distribution is achieved by controlling the speed of the pistons which apply the coating to the mandrel. The process, it is claimed, produces good neck mouldings and eliminates weld lines and waste.

Injection blow-moulding. In this process, a mandrel is coated with molten polymer in a split mould; the clearance between the mandrel and the inside of the mould cavity at any point determining the wall thickness in the 'parison'. The mandrel and its coating is removed from the mould, transferred to the blowing mould and inflated through the mandrel. The advantages claimed are similar to those for the Siemag process, also there is no freely suspended parison to sag and so polymers of high M.F.I. can be used. Further, many existing injection moulding machines can be adapted quite easily to accommodate blow-moulding tools. Although the process is gaining rapidly in popularity for certain applications, typical mould costs are higher than those associated with normal injection or extrusion blow-moulding.

5.5 FUTURE DEVELOPMENTS IN EXTRUSION PROCESSES

The arguments for and against the use of larger extruders or several smaller extruders are likely to continue, as are the discussions about the merits of single versus twin-screw machines.

Extruders and the extruder-based processes have for many years been subjected to mathematical analysis of varying degrees of accuracy and complexity. The design of extruder screws is a particularly fruitful field, although the machinery manufacturers claim, as do the car makers, that whereas you can have what you want if you are prepared to go to a specialist and pay for it, in most cases the standard range will fulfil most ordinary requirements.

The development of larger and faster computers and the use of more sophisticated methods of calculation now means that there is a better chance of setting up and solving more meaningful models of the various extrusion processes. This, it is hoped, will make it possible to produce more automated lines capable of operating closer to the limits of efficiency.

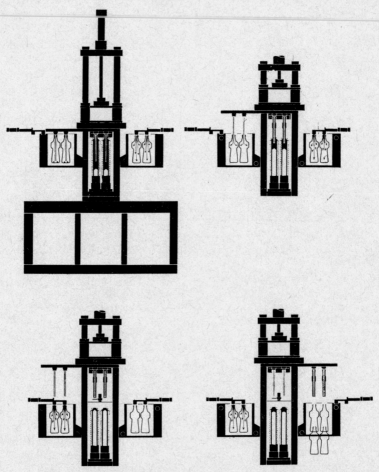

FIG. 5.43. The four phases of the Siemag dipping blow moulding process. (*top left*) Two mandrels in the dipping barrels starting the coating phase, blow starting on the twin left hand moulds and hold blow being applied to the right hand mouldings. (*top right*) The pre-forms almost complete on the dipping mandrels, forming blow almost complete on the left and hold blow still applied on the right. (*bottom left*) The pre-forms complete and the pistons fully up, hold blow being applied to the left and blowing finished on the right. (*bottom right*) Hold blow on the left, the mandrels in the centre about to enter the dipping barrels and the finished mouldings on the right falling clear to allow the pre-forms above to be dropped in on the mandrels. (Courtesy Siemag Siegener Maschinenbau GmbH, Germany.)

There is also likely to be an increased tendency for the machinery builders to supply complete extrusion lines designed as an integrated unit rather than an assembly of components.

Finally, of course, it seems likely that more attention will be given to the problems of re-using scrap generated by a particular process, and also to the reclamation and re-use of waste materials.

5.6 CALENDERING

5.6.1 Introduction

Calendering is a process used for the manufacture of continuous plastics sheet. The calender itself consists essentially of a complex of three or more

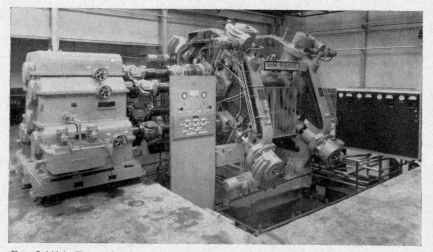

FIG. 5.44(a). Two calenders each with different roll assemblies. (*Note:* Independent drives to each roll, provision for roll pre-loading, automatic nip adjustment, and roll bending—also lubrication system.) (a) 26 in (0·66 m) × 72 in (1·83 m) Inclined 'Z' calender. (Courtesy Francis Shaw Co. Ltd., Manchester.)

heated rolls whose function is to convert the high viscosity polymer mass into film or sheet. Figure 5.44 shows a 4-roll inclined 'Z' plastics sheeting calender together with a diagram of the roll arrangement. The plastic mass is fed into the nip between the first two rolls whence it emerges as a film. It then passes round the remaining rolls, each of which has a particular function to perform in the calendering operation. In a typical 4-roll set-up there are three nips; the first controls the feed rate, the second acts as a metering unit. The thickness of the sheet is determined by the nip between the last pair of rolls and the surface by the last roll, which may be glossy, matt or embossed. Calendering has often been likened to extrusion, and it is easy to appreciate that this is so if the last pair of rolls is thought of as a die with rotating lips.

Fig. 5.44(b). 36 in (0·91 m) × 101 in (2·57 m) 4-roll 'L' calender. (Courtesy Farrel Bridge Ltd., Rochdale.)

Film and sheet can in principle be made equally well by calendering or by extrusion, although since the advantages and disadvantages of each process are not easy to define it is usually necessary to judge each case on its specific merits. However, in general we may say that polyethylene, polypropylene and polystyrene film and sheet are normally made by extrusion. In the case of PVC film and sheet, as well as for rubber processing, calendering is almost always used since the process is less likely to cause degradation than is extrusion.

Generally speaking, almost identical pre-forming as well as haul-off or post-forming equipment is used in production lines whether calenders or extruders are installed. For a given production tonnage it is usually necessary to have more extruders than calenders, and this is a situation which makes the extrusion set-up more flexible and more able to handle short production runs. However, calenders are now appearing which are capable of higher production speeds than are normally obtainable with extruders, and this again favours long runs with calenders.

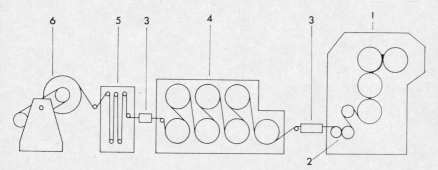

FIG. 5.45. Inverted 'L' calender plant for the production of plastics sheeting. 1. Calender. 2. Embossing calender. 3. Thickness gauges. 4. Water-cooled cooling train. 5. Wind-up accumulator. 6. Wind-up station.

5.6.2 The process

The calender is simply a piece of forming equipment and by itself is not capable of producing saleable material. It forms part of a complex line of ancillary equipment, the components of which are dictated by the materials being processed and the nature of the finished product.

The composition of a typical calender line for the production of plastics sheeting is shown diagrammatically in Fig. 5.45. The function of the mixing stage is to provide feed-stock in a form most conveniently handled by the calender. This may be as a simple hot melt in the case of, e.g. polyethylene. With PVC, however, a two-stage preparation process is commonly used. The first stage consists of pre-mixing at temperatures of up to 80°C. After this there follows a gelation stage. Gelation of flexible and plasticised compounds is carried out in an internal mixer of the Banbury type, followed by either 2 mills or 1 mill and an extruder. In both cases the material is fed into the calender as a continuous strip.

After leaving the calender the sheet passes over a stripper roll whose main function is to remove the hot material evenly from the calender rolls. Stripper rolls can be heated or cooled and in most cases form part of the cooling train although they may be used as booster heating devices where embossing is carried out.

The most widely used method of cooling the sheet is to use a sequence of water-cooled rolls although in some cases it is convenient to pass thick sheet (greater than 3·2 mm) directly into a water bath before winding up.

Although all the operations in the calendering sequence are important, it is vital to ensure close control of speeds and temperatures, since it is during cooling that the development of internal strains in the material is controlled.

A very large percentage of calendered sheet produced has a matt finish. Apart from the attractive surface appearance, the matt finish is functional. If polished sheets are of flexible PVC and are stacked for panel cutting, they stick together (block) to such an extent as to make the separation of the individual panels time consuming. A matt surface finish prevents this and also hides any scratch marks on the roll surfaces.

5.6.3 The calender

The development of calenders can be traced from the machines used for the various processes in the finishing of textiles in the 18th century. Leaving aside the two-roll mill, which is basically a mixer, the early rubber calenders date from the first half of the 19th century. Except for minor variations, the basic concept and design of these machines remained virtually unchanged for almost the next hundred years. There are several reasons for this long period of stability: firstly, the considerable size and cost of the equipment; secondly, the high engineering standards and the overall simplicity of the process made maintenance and renovating relatively easy; and thirdly, the comparative stability of materials and applications made it uneconomical and unnecessary to change.

By the 1930s, however, the gradual introduction of new materials such as PVC and the increasing pressure to raise standards of performance on products—for example, tyres, belting, etc.—made it necessary to design machines with improved methods of temperature and thickness control.

The Troester Company had developed in Germany a machine by the early 1940s which embodied separate drives to each roll, power-operated adjustment of the nip and cross-axis setting and flood lubrication. This calender was also capable of operating at temperatures in the region of 200°C. All of these features are used in machines of the present day.

It is clear from these remarks that the production of good sheet depends

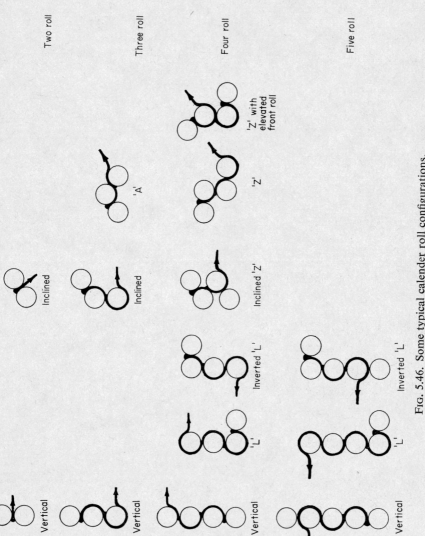

FIG. 5.46. Some typical calender roll configurations.

on a number of factors: (a) the rolls; (b) the control of temperature in the rolls; (c) the ability to set and maintain an accurate nip between rolls; (d) the drive system and lubrication. We can now, therefore, deal with each of these topics in turn.

(a) Calender rolls

It should be obvious that the most important components of the calender are the calender rolls since, by their surface finish, stability, and uniformity of heating, they control the final product. Calender rolls are normally made of chilled cast iron. Shock cooling suppresses the formation of graphite in the chilled region and encourages the production of very hard iron carbide.

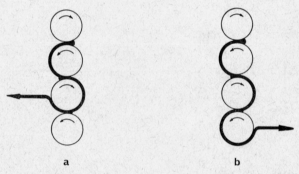

a   b

Fig. 5.47. Alternative feed and sheet paths for 4-roll vertical calender.

There are many ways in which the rolls in a calender can be combined and a selection of the currently used roll configurations is shown in Fig. 5.46. The selection of the appropriate roll configuration for any application will involve a number of factors which includes the nature of the material to be processed, sheet width, thickness, output rate, and so on. These considerations, coupled with whether the material is to be produced as an unsupported sheet or is to be combined with another ply as a laminate, will also govern the path of the material through the rolls. Figure 5.47(a) and (b) show different paths through a vertical arrangement, and Figs. 5.48(a) and (b) illustrate two different paths through four roll 'Z' calenders.

Early calenders almost invariably used the vertical stack arrangement. Offset designs such as the four-roll inverted 'L' and 'Z' types are now preferred since it is easier to feed stock into a horizontal nip, and also because such machines are much more compact. In addition, the problems of roll-float are minimised. Construction is also easier for the 'Z' types since less compensation has to be made for roll bending as there are no more than

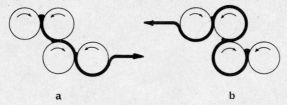

FIG. 5.48. Alternative feed and sheet paths for 'Z' type calenders.

two rolls in any vertical line compared with three in the four roll inverted 'L' and four in the vertical stack. The 'Z' configuration also has the advantage of a shorter sheet path with lower heat loss in the material.

(b)   Control of roll temperature

As has already been mentioned the quality of the finished product is directly dependent on the close control of temperature throughout the process. More specifically also, the nature of the product will dictate degree

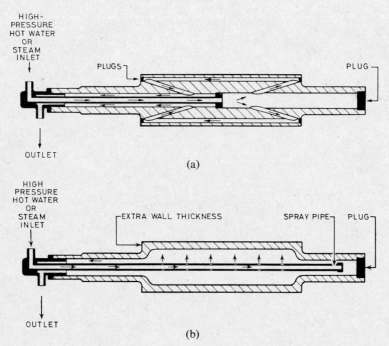

FIG. 5.49. Illustration of the methods of distributing coolant in calender rolls. (a) Drilled roll. (b) Cored roll. (Courtesy Butterworths.)

of temperature control over the working face of the roll; this is generally within $\pm 1°C$ up to a temperature of $200°C$ under normal working conditions. This uniformity is achieved by leading high pressure hot water or steam through the rolls and by circulating hot oil through the bearings in order to minimise heat losses at the ends of the rolls. Two main types of roll design are available (Fig. 5.49)—cored, and drilled. The former can give rise to temperature gradients but is adequate for some plastics and many rubber applications. Where greater accuracy is required though, drilled rolls are used. The heating system is generally interlocked with the roll drive so that distortion can be minimised by ensuring that the rolls are rotating before heat is applied.

(c)   Gauge control

During processing calender rolls are subjected to very high forces in order to squeeze the plastic material into a thin film. These forces cause the rolls to bend, thus giving film which is thicker in the middle than at the edges. This is counteracted in a number of ways, three of which are normally used commercially:

   (i) the use of 'crowned' rolls, i.e. rolls having a greater diameter at the middle than at the edges;
   (ii) gauge control by roll bending, in which an extra outer bearing is fitted to each roll. These bearings are connected to hydraulic systems which can cause the rolls to deflect. Although a useful method of adjustment, the amount of deflection obtainable is usually limited to less than 10^{-4} m;
   (iii) roll crossing: this method, which consists in setting the axis of one roll out of line with its neighbour, provides a method of increasing the camber effect. It operates by an increase in the nip opening at the end of the rolls, hence it is only a 'one-way' adjustment. Small local connections to roll profiles can also be made by the use of suitably placed induction heaters.

Gauge thickness is often monitered continuously in both the machine and transverse directions and the necessary variations fed directly into the roll control system.

(d)   Drive systems and lubrication

Three main types of calender drives are normally available:

   (i) single motor drive through reduction gear, and wheel and pinion mounted individually on roll extensions;

(ii) single motor drives through unit gear box and universal couplings to the roll extensions;

(iii) multi-motor drive through gear box and universal couplings to roll extensions.

Whatever method is used the motors must be capable of infinite speed variations (usually in the operating range 10–150 m min^{-1}) and this is generally achieved by a D.C. motor powered by a D.C. generator driven by an A.C. motor.

Lubrication is achieved by a flood system, into which two safeguards are built: (i) a fall in oil pressure causes the calender to stop; (ii) an interlocking system makes it impossible to start the calender until the oil pump is operating at the correct pressure.

5.6.4 Calender operation

The rate at which saleable products can be manufactured depends on a number of factors. These include: (i) the mixing and fluxing capacity in the line; (ii) the ways in which the calendering process affects the properties of the materials during processing; (iii) the quality of the product required—particularly with surface finish.

The type of product will also have an effect on the productivity of the line. For instance, heavy gauge sheeting ($2\cdot5 \times 10^{-4}$ m and over) is quite easy to make and run at speeds in excess of 60 m min^{-1}. Greater speeds can be achieved if this film is to be 'post-treated' by embossing, lamination, coating or printing, since defects can be masked by the post-calendering stage.

Thin films are often accompanied by layflat problems, although even here speeds of 100 m min^{-1} are common. Where really high gloss, thin, rigid sheets are needed, this figure may drop by a factor of 10.

As we have seen, the calender process is capable of manufacturing a very wide range of products each with its own specific processing sequence and attendant variables. It is, therefore, only possible here to deal briefly with a few of the main considerations which apply generally.

(a)   Feed methods

The hot gelled plastic is usually fed into the calender nip after passing through a metal detector—either as a continuous strip or as a succession of small strips cut from the mill and rolled into 'pigs' or 'dollies'.

Alternatively, an extruder which receives stock either from a mill or from an internal mixer can deliver a rod or tape directly into the calender nip. This latter method has the advantage that the material is passed through screen packs in the extruder, and also undergoes an additional homogenisation stage. Care must be taken, however, to avoid an excessive amount of heat build-up in the extruder, which should have a low L/D ratio.

Some attempts have been made to use a powder pre-mix as the feed-stock but it is not known whether this method is economically as well as technologically feasible.

(b)   Running the unit

One of the prime essentials of successful calendering is stability of operating conditions. This means that all changes have to be carried out gradually. At a roll heating rate of $1°C$ min^{-1} this can mean as long as 5–10 hr before the rolls and side frames are in equilibrium at the correct temperatures. The material too presents some heating problem, since the work heat is capable of raising roll temperatures by as much as $20°C$. This may necessitate a further settling down period of several hours after introducing the material until sheet of uniform quality is being produced.

The manufacturing programme should therefore be designed around long runs, and for very large production it is often more economical to have a battery of calenders each set to a particular product thickness rather than to attempt to achieve too great a degree of flexibility by adjustment.

In order to achieve an acceptable profit margin in calendering the interplay of the following variables must be taken into account:

(i)   raw material costs;
(ii)   labour costs;
(iii)   selling charges;
(iv)   size of plant and depreciation rate;
(v)   fixed and variable overheads.

(c)   Calendering faults

As with all other manufacturing processes the products of calenders are liable to suffer from a variety of faults, due in the main to one or more of the following causes:

(i)   formulation errors;
(ii)   inadequate compounding at the pre-mixing stage;
(iii)   calendering errors.

5.6.5   Variations on the process

Recently interest has increased in a compromise between extrusion and calendering. A two-roll calender is mounted at the discharge end of a conventional plastics extruder fitted with a simple spreader die. Although more expensive than an extruder with normal adjustable die lips, this set-up is cheaper than a calender set-up of equivalent output. Productivity is high since little time is lost in coming up to temperature, and gauge adjustments are easily and quickly carried out.

There are also advantages in taking the hot melt directly from the die into the nip. These include: a shorter heat history and exposure to oxidation; reduced orientation; less risk of entrapped air; and clarity of embossing on thick films since the pattern is imparted at film forming temperatures.

5.7 FIBRES

5.7.1 Introduction

Fibres are an immensely important commodity. Just how important may be appreciated when it is realised that the total world production of fibres currently exceeds 2×10^7 tonnes per year. Of this figure about 10^7 tonnes per year is accounted for by the production of 'man-made' fibres. The term 'man-made' fibres is used to distinguish this class of material from the natural fibres made from such materials as wool, cotton, flax, silk, etc. Even man-made fibres though are split up into two categories: (a) regenerated fibres, whose base is a natural polymer which is modified to produce such materials as viscose rayon, cellulose acetate, etc.; (b) synthetic fibres, where the polymer chain is built up from small molecules, as is the case, for example, with polyesters (e.g. terylene, dacron, etc.) and the various nylons.

It is quite clear, therefore, that these materials should be discussed in this book on Polymer Conversion, and I have included them in this chapter for two reasons. In the first place, all man-made fibres, and at least one natural fibre—silk, are made by some kind of extrusion process using a die. In the second place, a special type of fibre, whose applications are increasing rapidly, is made by slitting an extruded sheet into a multiplicity of very thin strips.

5.7.2 Fibre manufacture

Within the sub-divisions already mentioned, the different man-made fibres are generally classified according to their chemical structure. For our purposes, however, we are more concerned with the manufacturing and processing aspects, and it is convenient to classify them from that point of view. Table 5.6 gives some information on the nature and properties of the more familiar fibre forming materials classified according to their method of manufacture.

TABLE 5.6

Characteristics of some fibre forming polymers

Method of manufacture	Fibre type	Chemical nature	Tenacity[a] (g decitex^{-1})	Trade names	Comments
1. Wet Spinning Solution extruded into coagulating bath	Viscose rayon	Regenerated cellulose	2·5-5	Sarille, Evlan, Vincel	All textile uses, especially fine fabrics. Relatively poor chemical resistance
	Alginate	Calcium alginate	1·5-2		Flameproof, also soluble in alkaline soap solutions. Used as soluble sock separators
	Casein and other proteins	Protein	0·8-1	Lanital, Aralac, Fibrolane, Merinova Andil, Vicara, Soybean	Same feel as wool, used with wool and as blend with other fibres to improve 'comfort'
	Acrylonitrile	Copolymers of acrylonitrile with various co-monomers	2-3	Dynel, Acrilan, Courtelle	Good chemical resistance, used either alone as knitwear, etc., or in blends with wool and cotton
	PVC	Vinyl chloride	1·6-2	PE, LC, PCU, Isovyl, Rhovyl, Fibravyl Thermovyl	Non-flammable, rot proof and good light resistance. Used in chemical plant, protective clothing and furnishings
	Vinyl alcohol	Vinyl alcohol and formaldehyde	3-6	Vinylon, Kuvalon	High chemical resistance and resistance to water. Used in heavy duty clothing and fishing nets. A water-soluble version is available
2. Dry Spinning Solution extruded into air or steam heated atmosphere	Cellulose acetate	Acetylated cellulose	1-5	Celanese, Estron, Aceta, Alon, Tricel, Arnel, Tenasco, Cordura	Wide variety of uses in clothing, etc., used alone or as blends with wool and cotton. Also used in high strength form for tyre cords

Vinyl	Copolymers of vinyl chloride and vinyl acetate	2–3	Vinyon, Vinyon HH	High chemical and water resistance. Wide use including industrial applications and fishing nets. Low melting point
Acrylic	Copolymers of acrylonitrile	2–3	Orlon, Pan, Dralon Acrilan, Courtelle Creslan, Zefran	High strength, highly resistant. Wide range of textile uses, either alone or in blends with wool and cotton
3. Melt Spinning Molten polymer extruded into air				
Nylon	Polyamides, lactams, etc.	4–9	Antron, Cadon, Perlon, Caprolan, Rilsan, Nomex	Uses depend on chemical structure. As a class nylons are strong, can be heat set and are chemically resistant. Nomex is flame resistant
Polyesters	Generally polyethylene terephthalate	3·5–8	Terylene, Dacron, Kodel Vycron, Fortrel, A-Tell	High strength, wear and chemical resistant. Can be heat set. Wide range of uses in textiles, alone or in blends
Saran	Copolymers of vinylidene chloride	2–2·5	Velon, Permalon, Tygan	High chemical resistance. Used in fabrics exposed to light and bacterial attack. Wide use in car upholstery
Polyethylene	Ethylene homo- and copolymers	2–3	Courlene, Marlex	Good chemical resistance, used for industrial clothing and car upholstery
Polypropylene	Propylene homo- and copolymers	7·5–8	Merkalon, Ulstron	Excellent chemical resistance and high strength. Used as ropes, 'artificial seaweed' carpets

[a] The mechanical properties of fibres and filaments are normally measured in terms of tenacity and extensibility. Tenacity is obtained by dividing the breaking load in g by the denier or decitex of the yarn. The denier is defined as the weight in g of 9 kilometres of yarn, and the newer decitex the weight of 10 km of yarn.

It will be seen that there are three main methods of manufacture: (a) wet spinning; (b) dry spinning; (c) melt spinning. Although different in detail, all three processes use the same basic technique; the solid fibre forming substance is made liquid either by solution or by melting, and it is then forced by means of a pump through a spinneret which contains a large number of very fine holes. On emerging from the spinneret the fibres are drawn down to an appropriate diameter. At the same time a setting process operates: in the case of (b) and (c) by cooling and solvent evaporation, or by simple cooling. Where a coagulating bath is used the setting mechanism may be one of simple precipitation into a non-solvent medium. Alternatively, the bath can perform an additional chemical operation. For example, in the case of alginate fibres, the use of a coagulating bath containing calcium chloride converts the sodium alginate into calcium alginate which is rather more stable and 'spins' better.

5.7.3 Processing

Many of the materials listed in Table 5.6 are 'strengthened' by drawing after extrusion; for example, the high strength nylons used in tyre cords possess a tenacity of up to 8 g decitex^{-1} with an elongation at break of 16–20%. In applications where softness of handle is more important, such as in clothing or furnishing fabrics, a representative value would be a tenacity of 3 or 4 g decitex^{-1} and 30% elongation. Figure 5.50 shows in schematic form the manufacture of both filament yarn and staple fibre.

Staple yarn production

Among the natural fibres only silk is produced in a continuous length, all the others—cotton, wool, flax, jute, etc.—occur in fibres ranging from 2 cm to more than 50 cm in length. These short lengths are converted into yarn by arranging them into roughly parallel bundles which are gradually thinned out by pulling and twisting. Over the years textile manufacturers have developed extremely complex high speed equipment to carry out the process of converting the fibres into 'staple' yarn as it is called, and it would seem logical that attempts should be made to achieve the properties of natural staple yarns with man-made materials. Apparently, in fact, the first attempts were made in the 1920s to combine shredded viscose rayon with wool in order to use up the considerable quantities of waste filament generated by the continuous production of viscose rayon. These early attempts were discouraging until a reliable method of making short uniform fibres was developed. Three general methods are now available: (a) wet cutting; (b) dry cutting; (c) continuous processing.

One of the main differences between the production of man-made fibres that are used as continuous filaments, and those which are destined for conversion into staple yarn, is that whereas the former are gathered from a spinneret containing relatively few orifices (about 50), the latter are often produced as tows containing many thousands of filaments.

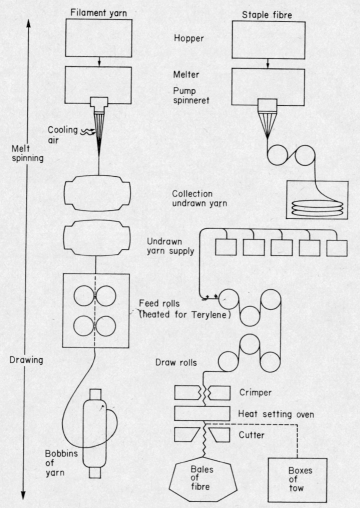

Fig. 5.50. The manufacture of filament yarn and staple fibre. (Courtesy I.C.I. Ltd.)

(a) *Wet cutting.* The rope or tow of filaments—which has already undergone some stretching—is cut into precise lengths which are then dried. Some shrinkage usually occurs which imparts a little crimp.

(*b*) *Dry cutting*. In this method the tow is dried before cutting which reduces the development of crimp.

Both of these methods result in random bundles of fibres which are then baled. The third method retains the parallel arrangement of the filaments.

(*c*) *Continuous cutting*. The tow is handled in such a way that although the filaments are cut at regular intervals, only a proportion of the bundle is cut at each interval and the tow is never completely severed. Three methods of severing the individual filaments are generally used: abrading, cutting and breaking.

Textured yarns

Many devices have been introduced over the years both with natural and man-made fibres to create novel effects. Most of the effects have been achieved by different amounts of twist, by sudden changes in thickness, or by the use of a combination of two or more different fibres. One development which is gaining popularity with great rapidity has been the introduction of stretch fabrics. These are made from the synthetic fibres which can be heat-set; the stretch characteristics of the yarn are generally put in by one of the following techniques:

(1) *Twisting:* the yarn is twisted and then heat-set, after which it is untwisted. Some commercial examples of this type of stretch yarn are Helanca, Fluflon, Superloft, Saaba and FT(ARCT).
(2) *Hot-Drawing:* the yarn is drawn over a heated edge which, since it heats only one side of the yarn, imparts a curl. This method gives a similar effect to that achieved naturally with wool, and a typical product is Agilon.
(3) *Stuffing Box Method:* here the fibres are forced into a chamber, crimped and heat-set. This process produces highly bulked yarns which are marketed under the names of Banlon and Spunize.

5.7.4 Fibres from film

As we saw in Chapter 1 orientation of the molecules in the plane of a plastics film can produce a material of enhanced strength (see Table 1.3). When the orientation is uniaxial, this increase in strength is often achieved at the expense of strength in the perpendicular direction. For example, polypropylene or polyethylene terephthalate film which has been highly oriented in the machine direction is so weak in the transverse direction that it shows a marked tendency to 'fibrillate' into a number of thin strips.

It was recognised a few years ago that if this process could be controlled it would provide an efficient and relatively simple method of producing fibres from film. At that time the only other method available was to slit wide film into narrow strips by means of rotating or stationary knives. The slitting equipment, which had to be capable of operating with high precision, was expensive and the film could not be slit into widths below 1 mm.

The fibrillation process, which finds its greatest success with polypropylene film, takes the extruded film which is then oriented in the machine direction and passes it over a rapidly rotating roller on which are mounted one or more 'combs' containing a large number of small sharp teeth. These teeth cause the film to split into thin strips (which can be less than 1 mm in width) and these are then collected together and handled in the same way as normal textile fibres.

In an alternative method, generally used for the production of tapes exceeding 2 mm in width, the film is first slit by passing over a series of razor blades mounted on a bar. Stretching, and sometimes relaxation, takes place after the slitting stage.

The fibres produced in this way are extremely strong and rot-proof and thus widely used for carpet backing, agricultural sacks and binder twine.

Although in principle the process is simple and works well with polypropylene, there are often problems when other polymers are used. With polyethylene terephthalate, for example, it is difficult to generate and maintain a uniform fast-travelling split, and there are also indications that the shape of additives such as pigments incorporated in the film has an effect on the successful operation of the fibrillation stage.

5.7.5 Non-woven fabrics

For many centuries fabrics have been made by weaving or knitting with single filaments or yarns made by twisting together short fibres. Although perfectly satisfactory for many purposes, fabrics made in this way suffer from a number of disadvantages. In the first place the twist required for yarn making reduces the strength of the yarn, and secondly the relatively sharp changes in direction due to the interlacing of the yarns during weaving or knitting tends further to weaken them. Also twisted yarns tend to pack closer together. This reduces the amount of air which can be trapped within the fabric, thus reducing its insulating efficiency.

Weaving and knitting both require reciprocal motion in the interlacing of the yarns and this tends to restrict the speeds which can be obtained with these processes. A process already exists for making fibrous structures without the necessity of making yarns or of combining them by mechanical interlacing. This process, of course, is paper-making in which short fibres are dispersed in water, deposited on a wire screen, and subsequently dried. It is a technique which has been successfully adapted for the production of non-woven fabrics where the fibres are dispersed with air rather than water, and is particularly convenient since the cost of drying off large volumes of water is a very significant part of the total cost of making paper.

At first the fibres were coated with an adhesive in order to give coherence to the fabric. More recently, however, non-woven fabrics have been made without adhesives by using bi-component fibres. These fibres emerge from the spinneret as coaxial extrusions consisting of a high melting core surrounded by a low melting sheath. After processing, when the short lengths of fibre are combined to form a mat, the temperature is raised to the point where the sheath polymer melts and joins the core fibres at the points of contact.

Fabrics made in this way are known as melded fabrics and a variety of properties can be achieved by mixing normal single component fibres with bi-component fibres.

Among the applications for these melded fabrics are floor coverings, upholstery, blankets, and more recently for overcoats which combine thickness and therefore warmth with lightness.

5.7.6 Handle, feel and appearance

The fabrics made from natural fibres each have their own distinctive 'handle' and 'feel'. Wool, for instance, feels warm and is light and springy. Pure silk has an attractive feel and drape and a characteristic rustle. Fresh linen feels crisp and clean.

These assessments are all subjective and describe the impact of the material on the senses. Nevertheless, the feel of a fabric to the user is often of greater importance than its other qualities.

Man-made fibres owe their characteristics to a number of factors. The chemical structure of the polymer will to a large extent govern the strength and stiffness of the fibre, as well as its resistance to water and chemical attack. Physical form is also important; whether it is stretched, crimped or twisted all contribute to the characteristics of the fabric.

One of the most important factors, however, in determining both the handle and appearance of the fabric is the cross-section of the filament. Most melt spun fibres have a remarkably perfect circular cross section. Filaments which have been spun from solution are circular in cross section on emerging from the spinneret but collapse and become irregular as solvent is lost in the dry spinning process, or under the influence of the contents of the coagulating bath. On the other hand, if wet spun fibres are stretched to a considerable extent during spinning, the cross-section becomes more nearly circular.

It is possible to make a number of generalisations on what contributes to handle and appearance of fabrics. For example, a fine filament is associated with softness. A highly twisted yarn gives a harsh fabric. Flat filaments also tend to produce harsh fabrics with a bright appearance where the light is reflected from the flat surfaces. For a pleasing appearance it is also essential to have filaments of uniform cross-section along the filament length. Cross section is also important, and Table 5.7 summarises the contribution made by some of the more familiar cross-sections, to the qualities of the fabric.

TABLE 5.7

The effect of filament cross-section on the feel of man-made fibres

Cross-section		
Circular	◯	Gentle handle and feel, but with poor covering power[a]
Lobed—almost circular		Attractive handle, relatively good covering power
Elliptical	◯	Little difference from circular
Flat	⬭	Harsh unpleasing handle; feels greasy
		High lustre and good covering power
Triangular	△	Handles like silk (which also approximates to triangular cross-section). Lustrous appearance
Trilobal		Generally similar to triangular, fabric has a glittering appearance

[a] Covering power is measured by winding the yarn onto a bobbin bearing a scribed surface. When the lines could just be seen no longer, the yarn was removed and weighed. A low weight was equivalent to a good covering power.

FURTHER READING

Brydson, J. A., and Peacock, D. G. (1973). *Principles of Plastics Extrusion*, Applied Science, London.

Eldon, R. A., and Swan, A. D. (1971). *Calendering of Plastics*, Butterworths, London.

Fenner, R. T. (1970). *Extruder Screw Design*, Butterworths, London.

Fisher, E. G. (1971). *Extrusion of Plastics*, Butterworths, London.

Jacobi, H. R. (1963). *Screw Extrusion of Plastics*, Iliffe, London.

Miles, D. C., and Briston, J. H. (1965). *Polymer Technology*, Temple Press, London.

Moncrief, R. W. (1970). *Man-made Fibres*, 5th edn., Butterworths, London.

Ogorkiewicz, R. M. (Ed.) (1969). *Thermoplastics*, Iliffe, London.

CHAPTER 6

SECONDARY PROCESSING AND FABRICATION METHODS

6.1. THERMOFORMING

6.1.1 Introduction

Plastics in sheet form can be moulded into articles by pressing the softened sheet into or round a mould. In the widest sense of the term thermoforming can be taken to include the moulding of both thermoplastic and thermosetting sheets as well as laminates, also the solid phase forming of components. These latter topics are dealt with elsewhere and here we shall consider only the manipulation of thermoplastic sheets.

6.1.2 Methods of forming

Although the simplest conception of thermoforming is the formation of articles such as domes and canopies by 'free-blowing', it is usual to divide thermoforming into three main methods:

(a) vacuum forming;
(b) pressure forming;
(c) matched moulds forming.

(a) Vacuum forming

Basically the process uses a sheet which is clamped in a frame and then heated. When the sheet is rubbery the heat is removed and a vacuum is drawn between it and the mould. The pressure of the atmosphere above the sheet forces it on to the mould where it cools and solidifies.

This method is capable of producing only simple designs since severe thinning usually occurs at the corners and the bottom of the cavity where a female mould is used. With a male mould the thickest section is at the top instead of at the side walls.

In order to minimise these deficiences and make possible the production of more complex shapes a number of variants of the basic technique are

TABLE 6.1
Methods used in vacuum and pressure forming

Method	Description of process	Points to note
Simple vacuum forming	Vacuum applied and sheet drawn into or over mould	Limited to simple, shallow design. Equipment can be simple and relatively cheap
Drape forming	Mould moves up into the sheet and vacuum is applied	High draw ratios are possible but equipment is relatively complex
Plug-assist vacuum forming	Sheet is 'pushed' partly into mould before vacuum is applied	Allows deep drawing, more even wall thickness and shorter cooling cycle. Close temperature control needed, also equipment complex
Snap-back vacuum forming	Sheet is drawn into box, the mould is then lowered, and the vacuum released allowing sheet to 'snap-back' onto the mould. A second vacuum can be applied to ensure better contact with mould	The application of the second vacuum allows complex parts with recesses to be formed
Pressure snap-back vacuum forming	Similar to drape forming; the reverse of snap-back vacuum forming	Suitable for deep drawing but equipment is expensive
Plug and ring vacuum forming	Plug shaped mould introduced into sheet. A ring of size appropriate to the mould is incorporated in the clamp	Very simple technique, used for shallow drawing of tough sheets
Blow back forming	Similar to plug and ring vacuum forming, but pressure is used to free sheet	Suitable for deep drawn parts in thick, tough sheeting since high pressure can be used

used, and the essentials are summarised in Table 6.1. The details of these methods are shown diagrammatically in Figs. 6.1, 6.2, 6.3, 6.4, 6.5, 6.6 and 6.7.

(b) Pressure forming

This technique is essentially the same as vacuum forming except that an applied positive air pressure forces the sheet onto the mould. Since pressures greater than atmospheric can be used the process is capable of handling heavier, tougher materials, and producing more complex mouldings.

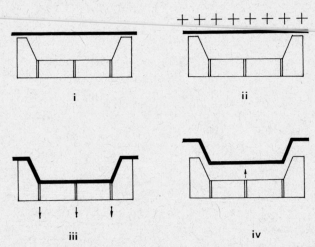

FIG. 6.1(a). The vacuum forming process using a female mould: (i) clamping stage; (ii) heating stage; (iii) vacuum applied; (iv) removal of formed article.

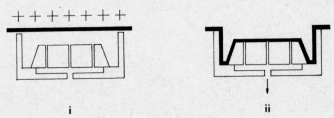

FIG. 6.1(b). The vacuum forming process using a male mould: (i) heating stage; (ii) moulding stage.

(c)   Matched moulds forming

This method is similar to the metal stamping process; the heated sheet is formed by pressing it between a pair of matched male and female moulds. This process is also used with foamed and laminated materials.

6.1.3   Thermoforming materials

Most thermoplastics can be thermoformed successfully although the ideal candidate needs a number of characteristics which are specific to the

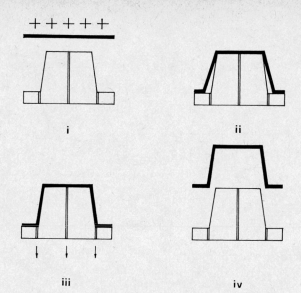

FIG. 6.2. Drape forming: (i) heating stage; (ii) draping; (iii) vacuum applied; (iv) removal of formed article.

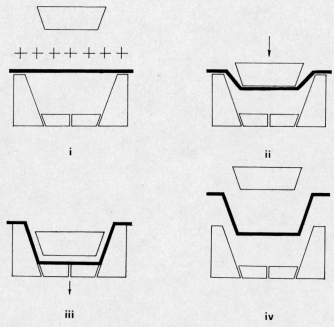

FIG. 6.3. Plug-assist vacuum forming: (i) heating stage; (ii) plug 'forming'; (iii) drawing with vacuum applied; (iv) removal of formed article.

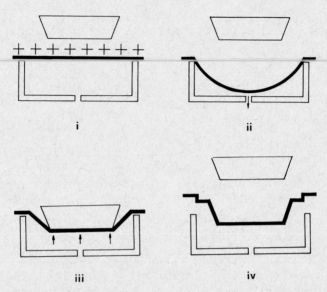

FIG. 6.4. Vacuum snap-back forming: (i) heating stage; (ii) vacuum applied; (iii) snap-back; (iv) removal of formed article.

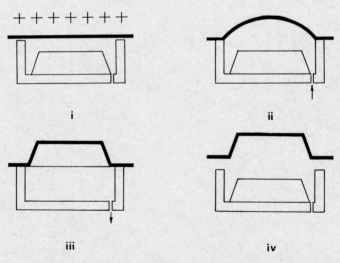

FIG. 6.5. Pressure snap-back vacuum forming: (i) heating stage; (ii) air blowing; (iii) vacuum forming; (iv) removal of formed article.

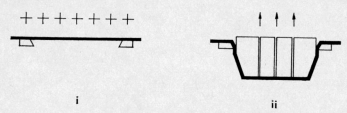

FIG. 6.6. Plug and ring vacuum forming: (i) heating stage; (ii) application of vacuum.

thermoforming process quite apart from the demands of the final product; such as cheapness, good weather resistance (for signs and other outdoor applications), high impact strength, etc. These desirable factors include: (a) low specific heat—for rapid heating and cooling; (b) high thermal conductivity—to enable thick sheets to be easily handled; (c) a 'formable range' as far as possible below the softening point—to prevent sagging in the heating frame.

Many materials do not possess these qualities (see Table 6.2) but their performance is improved by the use of such devices as assisted cooling by air blasts, water-sprays and choice of a high conductivity mould material. Heating can be made more effective by the use of heating elements on both sides of the sheet; these elements are also frequently wired up so that differential heating can be used across the sheet to ensure greater control of the wall thickness of the moulded article.

Each plastic has a temperature range within which it can most readily be formed and the choice of the forming temperature within the range is

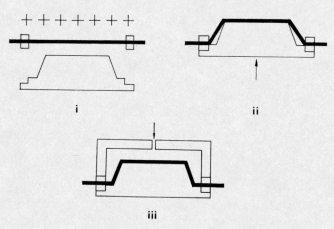

FIG. 6.7. Blow back forming: (i) heating stage; (ii) initial forming by the mould; (iii) application of air pressure.

TABLE 6.2

Materials for thermoforming

Class of polymer	Material	Thermoforming characteristics
Vinyl polymers	Polystyrene	G.P. sheet is brittle, but biaxial orientation produces better and clearer material. Toughened polystyrene is cheap, easily formed with good depth of draw, but has poor gloss and bad weathering properties. Expanded polystyrene is preferably formed with matched moulds particularly where deep draws or complicated shapes are required
	ABS	Good, tougher than high impact polystyrene
	PVC	Easily formed, has good clarity and is used where high chemical resistance is needed
Acrylic polymers	Polymethyl-methacrylate	Rather expensive, easily formed with excellent clarity and weathering characteristics
Polyolefins	Polyethylenes (LD and HD)	Difficult to use because of high specific heat, low thermal conductivity and relatively high crystallinity

dependent upon a number of variables. For example, the hot strength of the material falls off with sheet temperature, therefore, it would appear desirable to form at as low a temperature as possible. However, since the energy required to stretch the sheet is high at low sheet temperatures, these two factors conflict and a compromise must be made. Also the level of internal stress in the moulding is highest when forming takes place at low temperatures.

6.1.4 Thermoforming in practice

One of the main advantages of thermoforming is that moulds are relatively inexpensive. For prototype work and short runs such materials as wood, plaster and metal-filled epoxy resins are commonly used. On larger production runs moulds are normally made from steel or aluminium alloy.

Thermoformed articles can be made easily and cheaply with relatively simple equipment using rolls of film or sheet. Figure 6.8 shows, for example, equipment used in the production of high impact polystyrene containers. A point worth noting, incidentally, is the very deep draw obtained with this set-up. However, the thermoforming process is often

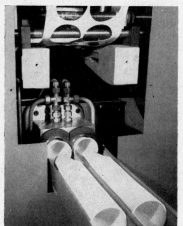

FIG. 6.8. The Illig RDKM, high speed forming machine. (*top*) General view of the machine. (*bottom left*) Forming area and punched web of material. (*bottom right*) Examples of deep drawn containers made on the machine. (Courtesy Adolf Illig Maschinenbau, Heilbronn.)

best carried out so that moulding and filling with the product to be packed are performed in the same production line. Where high speed, high volume production is concerned, an extrusion line is frequently coupled to provide sheet for the process.

Obviously, all requirements are different and a thorough and careful study must be carried out before investing in thermoforming equipment. The study must take into account such factors as:

(a) nature of the product;
(b) volume of production;
(c) cost structure of the product-range of products;
(d) likely modifications to the system.

After this has been done it is necessary to consider such factors as:

(i) the degree of automation envisaged—materials handling systems, in-line film extrusion set-up;
(ii) the type of forming technique to be used (e.g. see Table 6.1);
(iii) the degree of flexibility which is needed in the layout;
(iv) the amount and type of ancillary equipment—printing and decoration, filling, trimming, etc.

As an example of the type of layout that is involved in the production of one relatively simple product, Fig. 6.9 shows a representation of a line capable of producing polystyrene dairy cups at a rate of 1000 million per year.

6.1.5 Miscellaneous forming techniques

From time to time processes are developed which combine a number of different manufacturing techniques in one machine. Such an invention is the Topformer recently introduced by Daniels Hamilton. This process, which is shown diagrammatically in Fig. 6.10, combines injection, compression and thermoforming in one compact unit. Since the system converts granules into finished containers without the need for intermediary sheet production, the problems of indexing, pre-heating, punching and disposal of the offcuts are eliminated. Also, by careful design of the cross-section of the moulded disc it is possible to control the wall-thickness of the finished container to a greater extent than by using extruded sheet.

As we have already mentioned, much of the thermoforming production is of the 'form and fill' type. There are a number of variations of the process and those most commonly used are as follows:

(a) Matched-half containers. Here the two halves of the container—often from the same sheet with the joining web used as a hinge—are formed simultaneously, and subsequently cemented or welded together before filling.

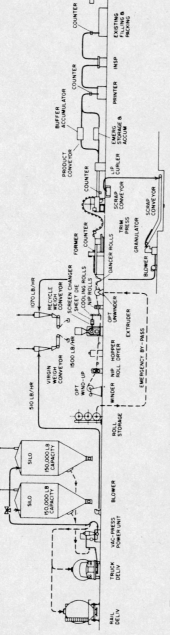

Fig. 6.9. Sketch of an automated line capable of producing polystyrene dairy cups at outputs of 10^9 per year. (Courtesy Processing Handbook.)

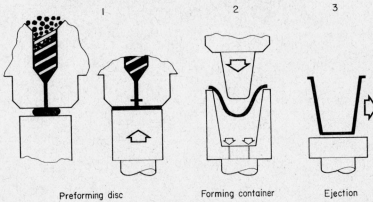

Preforming disc   Forming container   Ejection

FIG. 6.10. (*top*) Moulding thermoplastics containers. The three stages (*bottom*) run concurrently: in the first the disc is formed. The second stage consists of a type of plug-assist forming, and at the third stage the finished container is ejected. (Courtesy Daniels Hamilton Ltd., Stroud.)

(b) In the most familiar method the container is thermoformed, filled, and then closed by heat sealing on to it a film which may consist of several laminated plies or a single thickness of film. Often a simple snap-on closure is sufficient, but when diaphragm and closure are used together, the result is a neat pilfer-proof reclosable pack.

(c) The blister pack. This, in one of its forms, is very similar to the method just described; the 'blister' which contains the product is sealed

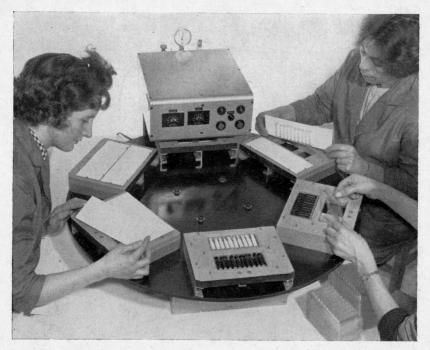

Fig. 6.11. Illustration of the sequence of events in blister packing. (Courtesy Sharp Intermatic Ltd., Aylesham, Kent.)

to a card which provides both a closure and a means of advertising or a set of instructions. Alternatively, the article can be mounted on a card and the film formed over it and on to the card. This technique in its turn is reminiscent of a form of overwrapping (see Figs. 6.11 and 6.12).

In the techniques described so far, with the exception of the free-blowing process, the finished article is produced by pressing the hot sheet against the correct sized male or female mould.

It is obviously possible, however, to use a combination of heating and shaping techniques, particularly with materials like acrylic sheet, to achieve the desired configuration.

FIG. 6.12. Illustration of a sleeve wrapping machine using shrinkable polyethylene film. (Courtesy Lerner Machine Co. Ltd., Enfield, Middx.)

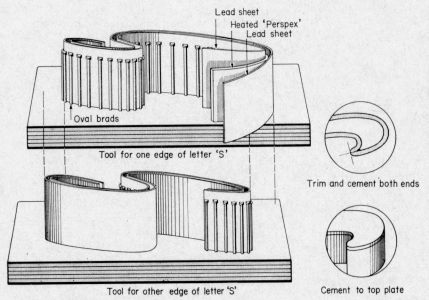

FIG. 6.13. Method of shaping a hot perspex strip to a complex single curvature. (Courtesy I.C.I. Ltd., Plastics Division, Welwyn Garden City, Herts.)

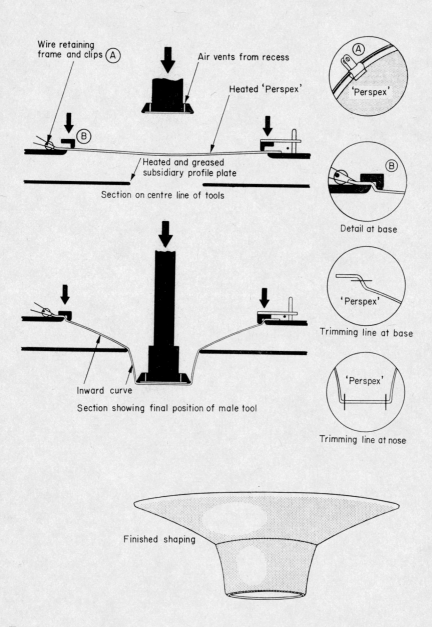

Fig. 6.14. Shaping acrylic sheet with a plunger and profile plate. (Courtesy I.C.I. Ltd., Plastics Division, Welwyn Garden City, Herts.)

Intricate shapes, as in this example of the script letter (Fig. 6.13), are produced by bending the heated sheet in contact with strips of lead and confining it against rows of nails.

The light fitting base (Fig. 6.14) is made by pressing with a plunger through a profile plate, and may be likened to a simplified version of plug and ring vacuum forming (e.g. Fig. 6.6).

The method is simple and efficient, and produces uniform thickness distribution. This is important so as to ensure that the light transmission characteristics do not suddenly change over the surface of the fitting. The more familiar methods of using a male or female mould not only would show relatively wide thickness variations, but would result in one or other of the faces being marred by contact with the mould surface.

6.2 FABRICATION TECHNIQUES

6.2.1 Introduction

We have seen how polymeric materials can be formed into shapes by the techniques of extrusion, moulding, thermoforming, etc. For some purposes these techniques are sufficient in themselves to produce the finished article. There are instances, however, when several assembly steps are involved and components have to be welded, stuck or otherwise joined together. Alternatively, it may be convenient, particularly on short runs or one-off jobs, to use different methods of manufacture such as machining, or solid phase forming. Indeed, unless very large production runs are contemplated, machining small components may often be cheaper than injection moulding as can be seen from Fig. 6.15. Another advantage offered by machining over injection moulding is that design modifications can be introduced during production relatively easily and cheaply. Hand lay-up forming methods are also extensively used, but these are described elsewhere (see Chapter 7).

6.2.2 Machining

Many of the more rigid plastics materials such as the acrylic polymers, polystyrene, polyacetal and the polycarbonates can quite easily be machined by the conventional techniques of cutting, blanking, drilling, turning, milling and routing. Components can also be finished by the usual methods of grinding, scraping, buffing and polishing.

Machining and finishing techniques

Although the standard metal and wood-working techniques can be used on plastics with very little modification, considerable care has to be taken to avoid overheating the work. In particular, compared with metals, acrylic plastics have very low thermal conductivity and softening points. At about 100°C the swarf will soften and become gummed to the piece being machined or to the tool; excessive heating can also cause crazing of the surface. For these reasons efficient cooling is important, but unlike the case of

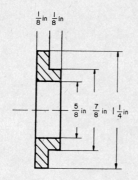

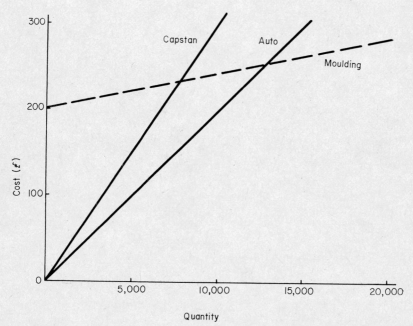

Fig. 6.15. Cost breakdown by process of producing the simple component illustrated above. (Courtesy A. Banham, Nylonic Engineering Co. Ltd., Rickmansworth, Herts.)

TABLE 6.3
Cubical coefficients of expansion of a range of materials

Material	Cubical coefficient of expansion (γ per $^{\circ}C$ at $20^{\circ}C \times 10^{5}$)
Glass (Pyrex)	0·9
Steel	3·2
Brass (66 Cu/34 Zn)	5·7
Aluminium	7·7
Polycarbonate	20
Polymethylmethacrylate	21
Polystyrene	21
Epoxy resin	21
Phenol formaldehyde resin	21
Polyacetal	24
Polypropylene	24
Polyethylene HD	36
Polyethylene LD	45

FIG. 6.16. Some optical components machined to a high degree of accuracy from acrylic polymer. (Courtesy Combined Optical Instruments Ltd., Slough.)

metals, in which the coolant is included primarily to protect the tool tip, in the machining of plastics it is the surface of the plastics material which must be cooled.

The coefficients of thermal expansion of plastics are also considerably higher than those of metals (see Table 6.3). Thus machined components should, if their dimensions are at all critical, be left for several hours after machining before measurements are taken.

For critical machining, when dimensional stability is important, residual stresses should be minimised by 'normalising' before and annealing after machining.

Some examples of optical components for a variety of industrial and scientific applications machined to close tolerances are shown in Fig. 6.16.

Machining is widely used in the fabrication of articles from PTFE. Compacted preforms are heated either with or without pressure to temperatures above 327°C so that the individual particles coalesce. After this the shaped mouldings can be machined to size.

These observations and fabrication techniques apply generally over the range of 'machinable' plastics, but many small but important differences exist between materials; for example, drill point angles differ considerably. When drilling acrylic materials the normal included angle of 118° is suitable for thick sections but about 140° is preferred for thin sheets. On the other hand with polystyrene an angle of 90–120° is usual for large or deep holes and 60–90° for small diameter and shallow holes.

6.2.3 Assembly methods

We have already seen that one of the principal benefits which arises from the use of plastics is that both the number of components and the number of finishing and assembly operations can often be significantly reduced. Another and particularly interesting example is the use of polyacetal components in the Astrolon watch by Tissot (Fig. 6.17). Not only has the number of components been reduced from 68 to 40, but since they are moulded in one piece, costly and time consuming finishing and assembly operations have been eliminated.

There are, however, many instances where components have to be joined together during the progression towards the finished article. The method used will depend on a number of factors which include:

(a) the properties of the materials involved;
(b) cost of the assembly process;
(c) where the assembly is to be carried out;
(d) whether or not the assembly is to be permanent.

The methods available, taking into account the factors we have just mentioned, are numerous, but for our purposes it is sufficient only to describe the techniques in general terms, and they fall conveniently into the following categories:

(i) welding;
(ii) cementing;
(iii) mechanical fastening;
(iv) press and snap fits.

(i) Welding

In order to join two components together by welding, heat must be supplied, with or without pressure, and the methods available are as follows.

Fig. 6.17. The use of polyacetal moulded parts in the Astrolon watch by Tissot. (Courtesy Hoechst A.G., Frankfurt.)

(a) *Heat-sealing.* Strictly speaking welding takes place when the two parts to be joined are melted by the application of heat and united. On the other hand, heat-sealing properly employs a heat-sealable coating on infusible or 'difficult-to-seal', e.g. biaxially oriented, films. However since the methods used for sealing and welding are similar we shall discuss them here.

The technique is generally used to adhere a thin to a thick piece of material—for example, a lid to a container—or two thin films together.

In order to heat up the material to be sealed to the temperature where fusion occurs, a certain amount of heat (Q) must be provided, the amount depending upon such factors as:

(i) the dimensions of the join (V);
(ii) the specific gravity of the film (γ);
(iii) the temperature difference (dT);
(iv) the specific heat of the film as a function of temperature ($C_p(T)$);
(v) the enthalpy difference of the material over the temperature difference (H).

These quantities are related as follows:

$$Q = V\gamma H$$

where
$$H = \int C_p(T)\,dT$$

Additional heat—depending on the method used—is lost by conduction through the different parts of the equipment.

TABLE 6.4

Characteristics and sealing performance of plastics films

Material	Characteristics	Performance
Polyethylene (LD) Polyethylene (HD)	High coefficient of thermal conductivity, degradation temperature is much higher than melting temperature	Rapid heat flow allows sealing tool to be very hot; local overheating is not important
Polypropylene	Low coefficient of thermal conductivity, high degradation temperature	Less easily sealed than the polyethylenes
PVC (Rigid)	Low coefficient of thermal conductivity, low degradation temperature	Poor contact welding performance. High tan δ allows good high-frequency welding

Table 6.4 shows some of the material characteristics which influence the sealing performance of films.

Heat is supplied in two ways as follows:
(a) by thermal contact—using permanently heated bars or wedges

(usually covered with glass fibre reinforced PTFE film to prevent sticking), impulse heated strips or hot wires;

(b) by the generation of heat within the weld itself using high frequency methods or by ultrasonic techniques.

Weld strength is usually improved—particularly with oriented films— by allowing the material to cool under pressure.

A summary of the methods used to heat seal a selection of the more commonly used packaging film is given in Table 6.5. Films which cannot easily be sealed can be provided with sealing media either by coating or laminating (cf. Chapter 5).

TABLE 6.5
Methods used for heat sealing packaging films

	Sealing Methods				
Material	*Thermal contact*	*Impulse sealing*	*Hot wire*	*High frequency*	*Ultra-sonic*
Polyethylene LD	*	*	*		
Polyethylene HD	*	*	*		*
Polypropylene (unstretched)	*	*			
Polypropylene (stretched)			*		
Polystyrene		*			
PVC (rigid)		*		*	
PVC (flexible)		*		*	*
Polyamide	*		*	*	
Polyester (stretched)			*		
Cellulose acetate				*	
Polyvinylidene chloride				*	

(*b*) *Hot plate or heated tool welding.*   The two surfaces to be joined are held against a heated plate or tool and then clamped together while solidification takes place. This method is conveniently used for joining plastics pipes together either by means of a butt joint or with an injection moulded coupling. Two variants of this process involve: (i) the use of an electrically heated wire in the joint; (ii) a metallic insert in the joint interface which is heated by radiation heating. In both cases the metal insert remains in the joint and good design is essential.

(*c*) *Hot gas welding.*   This method is extensively used where large scale fabrications are required. The sheets to be joined are bevelled and placed side by side. Then a filler rod of the same material is melted into the groove between the two components using a stream of hot gas produced by a flame or an electrically heated welding gun. Compressed air is sometimes used, but where there is danger of oxidation, nitrogen or some other inert gas must be used.

(*d*) *Friction welding.* This method is suitable for joining circular parts. One part is held in a jig and the other in a rotating chuck. By bringing the parts together, frictional heat is generated which causes the polymer to flow. The chuck, which is usually spring loaded, is then stopped and the two parts allowed to cool under pressure. In some instances the jig and chuck assembly can be simplified and problems of alignment minimised by oscillating one of the parts through a small arc instead of rotating it.

(*ii*) Cementing

There is a wide variety of cementing techniques available for joining polymers. The materials most frequently cemented are styrene, ABS, nylon and acrylic polymers. Engineering polymers which are more difficult to cement are the polyolefin's, polyacetals, polycarbonates and fluorocarbons.

The methods used generally fall into one of the following three categories: (a) chemicals, including hot melt adhesives; (b) polymers in solution; (c) solvents.

Several techniques have also been used to enhance bond strengths; these include ultrasonic curing of the adhesive, and pyrotechnic adhesives which cause the temperature of the interface to rise.

(*iii*) Mechanical fastening

Several techniques may be used for joining plastics to plastics and also to other materials. These include bolting, screwing and rivetting. With plastics two factors (a) high thermal expansion (Table 6.3) and (b) in some cases brittleness, must be allowed for. With the more brittle materials it is helpful to radius—and even polish—the edges of holes and slots. Also, when rivetting, a heated punch can be used to minimise the chance of fracture.

An alternative to using metal rivets is to mould rivet lugs on one of the components whilst their location holes are provided in the mating part. These rivet lugs can be deformed by ultrasonic energy using very small shaped horn tips. This method, which is illustrated in Fig. 6.18 produces tighter joints than those obtainable from conventional rivetting.

(*iv*) Press and snap fits

These valuable assembly methods are used with similar and dissimilar materials, and, since by their use it is possible to eliminate fastening components, they provide economies both in number of parts required and in the number of assembly steps. Some examples of engineering applications using polyacetal are shown in Fig. 6.19.

By increasing the included angle to 90°, snap fit assemblies can be made permanent (and pilfer-proof!) since the forces required to separate the parts become excessively high and, by suitable design, difficult to apply.

The use of press fits and snap fits in one assembly means that it is often possible to produce a joint that is both mechanically strong and rigid, and

HEAD FORM	STUD DIA.	HEAD DIA.	HEAD HEIGHT	CENTRE-CENTRE DIA.	STUD HEIGHT ABOVE PART BEFORE STAKING
STANDARD	d	2d	0.5d	d	1.6d
LOW PROFILE	d	1.5d	0.25d	0.75d	0.6d

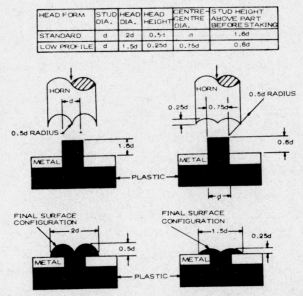

FIG. 6.18. The use of ultrasonic energy to deform rivets in joining plastics to metal. (Courtesy Morgan-Grampian Ltd.)

also leak proof. The design of joint and the amount of interference depends on a variety of factors, which include: size of component, mechanical properties (creep and tensile strength), coefficient of expansion and environmental conditions in service.

6.3 SOLID AND GAS PHASE FORMING OF POLYMERS

6.3.1 Introduction

We have already appreciated that some fabrication and finishing methods originally intended for use with metals can be used successfully with polymeric materials (see Sections 1.1 and 6.2). Other methods which have been used to advantage include compression sintering, forging and flame spraying. Although these techniques can be applied relatively easily to a number of polymers, they are often the only satisfactory way of converting

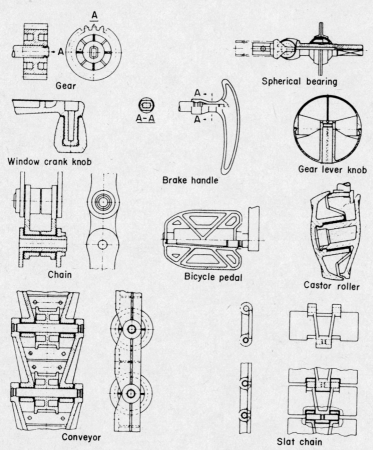

FIG. 6.19. Typical engineering design for acetal resin using snap fit assembly techniques to good advantage. (Courtesy Morgan-Grampian Ltd.)

some of the more thermally resistant polymers. One of the basic problems with high temperature polymers is the compromise between high temperature properties and ease of fabrication. For example, the polyimides are stable in air at 315°C, but fabrication is difficult and long moulding cycles are frequently needed. Modified 'easy moulding' grades are available but in each case the thermal resistance has been sacrificed. The thermoplastic aromatic polysulphones can be easily and rapidly processed but their thermal resistance is limited to 205–260°C.

One of the newer thermally stable polymers is the p-oxybenzoyl polymer known as EKONOL.† This material is highly crystalline but shows no melting point up to the decomposition temperature of 480–550°C. Fabrication must be carried out by metallurgical techniques.

The methods vary in detail according to the nature of the polymer and the product being manufactured, but the following techniques used with EKONOL are sufficiently representative to give as examples.

6.3.2   Compression sintering

Articles can be formed using normal compression moulding equipment in the temperature range 430–450°C and pressures of $3 \cdot 5$–14×10^7 N m^{-2}. Forming times from a few minutes up to an hour or so depend on the size and thickness of the sample. A typical moulding cycle is as follows: The material is preheated in an open mould at 260°C. After which it is then subjected to a pressure of about 3×10^6 N m^{-2} for 15 min in the temperature range 260–320°C. The pressure is raised to $1 \cdot 4 \times 10^8$ N m^{-2}, held for 2 min, and subsequently dropped to 7×10^6 N m^{-2}. The product is then allowed to cool at a rate of 35°C per hour.

There is in general considerable micro-porosity in compression sintered articles, but properties can be improved—and mouldings cheapened— by the addition of fine fillers.

6.3.3   Forging

Forgings have been made in a large number of different materials, including filled polypropylene, polyethylene, ABS, nylon and polyacetal.

Basically the technique involves heating a blank in an oven to a uniform temperature a few degrees below its melting point. The blank is then placed in a set of heated tools and forged in an hydraulic press. An illustration of the type of article which can be produced from relatively simple and inexpensive tooling is given in Fig. 6.20.

Plastics can be cold forged, but high pressures are required and, due to the viscoelastic nature of the material, considerable recovery is likely to occur. Hot forging is found to operate conveniently at pressures of about 6–$7 \cdot 5 \times 10^7$ N m^{-2} at speeds of 10^2 mm s^{-1} or so. Due to the poor thermal conductivity of most polymers, the preheating times tend to be rather long; a 20 mm thick polypropylene sample required a heating period of 1 hour at 162–164°C.

† A product of the Carborundum Company, Niagara Falls, N.Y.

An interesting property of forged components is the considerable increase in strength which accompanies increasing displacement in some polymers. For example, in the case of polypropylene, the maximum tensile strength of the material in the radical direction more than doubles with displacements above 40%. In the transverse direction the increase is smaller and in proportion to the level of displacement. Stiffness is also improved, but creep tests indicate that this improvement disappears when tests are carried out over long periods.

With a material such as EKONOL the preform is heated to about 175°C and the forming cycles are in the region of 6–10 sec. The energy generated during the forging which can be in the range $1\cdot5$–15×10^4 J is sufficient to produce flow and apparent fusion, since no microporosity is visible after forming. The method is also suitable for highly filled specimens since sufficient flow is achieved during forging.

6.3.4 Plasma flame spraying

A plasma of ionised helium is produced by passing the gas through a carbon arc, and the powdered material fed into the plasma. The effect of the very high temperatures ($\sim2750°C$) and pressures is sufficient to produce coherent coatings on a variety of substrates. The quality of such coatings is usually good, but can be improved by burnishing which eliminates pinholes.

6.4 DECORATION OF POLYMERS

6.4.1 Introduction

One of the principal advantages of using polymers is that, provided the forming process has been correctly designed, it is generally possible to use the component without subsequent finishing treatment. We have already discussed mechanical techniques (Section 6.2.2) including machining, abrading and polishing. Other manufacturing methods such as thermoforming blow-moulding, and some glass-fibre laminates require surplus material to be trimmed off, and mouldings occasionally need de-flashing. Often, however, a finishing process is used to impart a texture or colour to the product and one or more of the methods summarised in Table 6.6 may be used.

6.4.2 Methods of decoration

In addition to the techniques described in Table 6.6 two other methods are also available which find limited applications; these are engraving and two-colour processing.

(b)

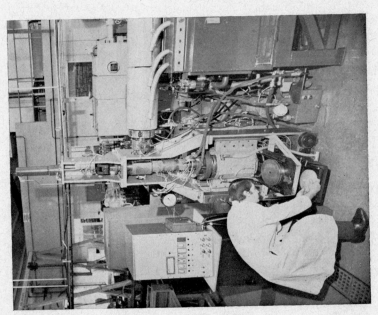

(a)

Fig. 6.20. Solid phase forming of plastics components. (*left*) The production of billets for solid phase forming. The machine is supplied with molten polymer by a conventional single screw extruder. (*right*) Component being removed from the die of the forging process. (Courtesy GKN Group Technological Centre, Wolverhampton.)

(c)

FIG. 6.20—*contd.* Typical components that can be made by the GKN polymer forging process. (Courtesy GKN Group Technological Centre, Wolverhampton.)

Several of the methods available for decorating plastics can achieve a similar effect, e.g. foil stamping, electroplating and vacuum metallising.

TABLE 6.6

Methods used in the decoration of plastics

Class of method	Variants	Observations
Adhesive work	Appliqué	Contrasting colours or textures cut out of similar materials and heat or adhesive bonded
	Labelling	Identification as well as decoration
	Hot foil application	Used in packaging and to 'plate' parts of mouldings, e.g. in automobile and domestic fields
	Transfers	Often used within the mould and permanently incorporated in the moulding
Surface coating	Painting and lacquering	Used for protective as well as decorative purposes. Clear and opaque lacquers are used. Typical examples are spraying plastics shoe heels to match the uppers
	Dyeing	The article is coloured by immersing in dyes, e.g. textiles and nylon mouldings
	Printing	Most of the standard printing methods are used and considerable ingenuity has been used in devising machines that will handle and print 'awkward' shaped containers. As with metals foil is often printed first and subsequently formed
Metallising	Electroplating	Success is very dependent on moulding history. Used widely with ABS for both decorative and electrical applications
	Vacuum metallising	Often used with a top coating of lacquer for protection

However, the equipment used and the cost of the process vary widely, and the choice of the method will depend on such factors as the material used, the type of article and the number to be decorated.

FURTHER READING

Bippus, W., and Ackerman, H. (1968). 'The Mechanical Fabrication of PVC Film and Sheeting in the Packaging Field', *Kunststoffe*, **58**, 197.

Bippus, W., and Heyse, K. *Methods and Equipment Used for the Welding and Heat Sealing of Packaging Films*, Kalle A.G., Wiesbaden-Biebrich.

Design Engineering Series (1970). *Plastics* (1), Morgan-Grampian, London.

'Perspex'—Machining—Shaping—Cementing', I.C.I., Plastics Division, Welwyn Garden City, Herts.

CHAPTER 7

MATERIALS IN COMBINATION

7.1 INTRODUCTION

Although the variety of polymers and their modifications are considerable, it is most unlikely that any single material will possess to the same extent all the properties needed for a particular application. Either—like polystyrene, for instance—it may possess good stiffness and transparency yet be a poor barrier to gases and vapours and have poor impact properties—or it may score highly in all these areas and may additionally possess high strength and flexibility, and yet, like polyethylene terephthalate, be difficult to heat-seal satisfactorily.

Some of these desirable properties can be achieved by the use of such techniques as copolymerisation or by changes in the arrangement of the molecular chains. However, the property improvements resulting from these techniques often only go part of the way towards satisfying the requirements of the application. An alternative method is to produce a composite structure in which several different materials are combined, each component being outstanding in at least one or more of the required characteristics. Thus, for example, we can improve the impact resistance of polystyrene by dispersing in it fragments of a rubbery material which act as stress absorbers. Alternatively, we can make aluminium foil into an attractive and efficient packaging material by coating it with a heat-sealable lacquer.

There are many ways of combining materials, as can be seen from the examples given in Fig. 7.1. Generally speaking, however, the range of composites can be covered by two familiar culinary analogies; the 'plum pudding' and the 'sandwich'. More scientifically, they are known as dispersion composites and laminates or layer structures.

In the case of the dispersion composite the alloying or reinforcing component (the disperse phase) is distributed throughout the material to be improved (the matrix).

Laminates or layer structures are those in which one material may be coated with another; or where several different materials are stuck together the achieve the properties which are needed.

The dividing line between these two types of composites is, as can be appreciated, hard to draw accurately and there must obviously be some overlap.

However, one simple way of looking at these two classes of materials and also of distinguishing between them is to visualise the dispersion

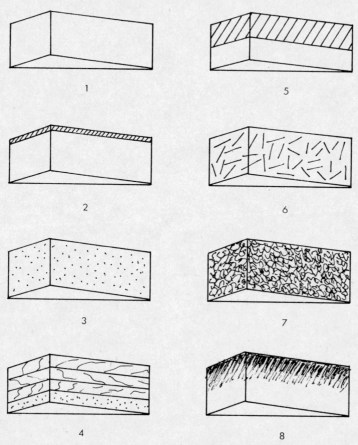

FIG. 7.1. Methods of making composites. 1. Alloying. 2. Coating. 3. Dispersion. 4. Laminating. 5. Bonding. 6. Fibre-reinforcing. 7. Powder-compacting. 8. Diffusion.

composites as consisting of a family of materials in which the matrix forms a continuous phase whilst the disperse phase is largely discontinuous. In the case of the layer structures we have two or more continuous phases generally with clearly defined interfacial planes.

There is, however, another group of important industrial materials

TABLE 7.1

Examples of typical resin/reinforcement combinations of high pressure thermoset laminates

Reinforcement	Phenol formaldehyde					Melamine formaldehyde					Polyester					Epoxy					Silicone				
Manufactured form	sheet	tube	rod	moulded macerated	moulded laminated	sheet	tube	rod	moulded macerated	moulded laminated	sheet	tube	rod	moulded macerated	moulded laminated	sheet	tube	rod	moulded macerated	moulded laminated	sheet	tube	rod	moulded macerated	moulded laminated
Cellulose paper	×	×	×	×	×											×	×								
Cotton fabric	×	×	×	×	×	×	×	×		×															
Asbestos paper	×	×	×																						
Asbestos fabric	×	×	×			×	×	×																	
Nylon fabric	×	×																							
Glass paper and mat										×					×	×									
Glass fabric	×	×	×	×		×	×	×		×	×	×			×	×	×	×		×	×	×	×		

roughly between these two categories. This is the family of laminates in which the disperse or modifying phase may be continuous but is also porous. This group subdivides also quite neatly into two categories:

(a) those in which the disperse phase is entirely surrounded or 'impregnated' by the matrix material;
(b) structures in which there is some interpenetration at the interface but where, in general, different materials are on either side of the boundary.

Category A comprises the laminates made by impregnating paper, woven cloth or a mat with a thermosetting resin. Typical examples are the hard wearing decorative laminates formed by pressure-bonding layers of tissue impregnated with phenolic or melamine formaldehyde resin. Table 7.1 gives a brief survey of the more familiar varieties. Where there are gaps in the Table does not necessarily mean that these combinations cannot and have not been made but simply that the ones indicated by crosses are the most suitable as far as physical properties and economic considerations are concerned.

Category B includes the 'leather-cloth' type of structure. Here a thermoplastic (or, less commonly, a thermoset) material is applied to one side of either a woven or non-woven substrate. Table 7.2 gives some examples of this type of structure together with applications.

Having dealt with this important 'no man's land' we must now consider in turn the two main categories of composites and see how their special characteristics and techniques of manufacture can help us to achieve the type of product we require.

TABLE 7.2

Some examples of 'leather-cloth' type laminates

Structure	Characteristics and uses
PVC/non-woven nylon	Made by melt spin-bonding, webs can be thermoformed and used in car and furniture upholstery (200% elongation at break)
Non-woven $\left\{ {nylon \atop polyester} \right\}$/GRP	Printed non-woven fabric gives 'finish' to boats, buildings, etc.
$\left. {PVC \atop Others} \right\}$/nylon fabric	Clothing, furnishings, etc.
$\left. {Thermoplastic \atop Thermoset} \right\}$/non-woven C fibre mat	High strength, high temperature and chemical resistance

7.2 DISPERSION COMPOSITES

7.2.1 Introduction

We defined this class of materials as one in which the alloying or reinforcing component is distributed throughout the material to be improved.
However, the differences between the so-called 'monolithic' and composite materials are often difficult to distinguish. For example, most 'crystalline' materials in normal use are in fact polycrystalline; Fig. 1.12 shows,
for instance, that a material like polyethylene consists of a crystalline
phase dispersed throughout an amorphous phase of the same material.
In its polycrystalline form graphite consists of disordered regions of carbon
atoms between the crystallites, and as an additional feature contains
minute pores, i.e. dispersed gas.

Steel also provides an interesting example of this problem of description since, depending on the amount of carbon present, we have a solid
solution of iron and carbon ($<0.8\%$ C), or a material which is strengthened by the presence of precipitated carbon (0.8–2.0% C) (cf. Section
1.3.3).

We can thus appreciate that the spectrum of materials which can be
created is not only very large, but capable of possessing almost infinitely
graded properties. Here we shall restrict ourselves to a consideration of
the characteristics of solid composites made by physically combining two
or more existing materials to produce a multiphase system with different
physical properties from those of the starting materials. There are cases
where some chemical change occurs during the formation of composites,
i.e. in the reinforcement of polystyrene by rubbers or the incorporation of
carbon black in elastomers (cf. Section 3.3.2). Generally, however, we
would expect there to be little alteration in the phases except perhaps to a
minimal extent at the interface.

Having described very briefly what we mean by a dispersion composite
it now remains for us to do two things: (a) discuss the extent to which the
modifying phase can alter the characteristics of the matrix and create
new properties in the composite, and (b) consider how the various types of
dispersion composites are converted into the finished article.

7.2.2 Mechanical properties

The range of dispersion composites based on polymeric materials is large
since the disperse phase varies not only in its chemical nature, from gases
to glass, but also in its physical form—from microspheres to long filaments.
The extent of this range is shown in Table 7.3.

From the standpoint of the converter to fabricator it is usual to divide
dispersion composites into two separate categories: (a) particulate com-

posites; and (b) fibre composites. This is appropriate since not only are the applications different for the two categories but so also are most of the methods of fabrication; particulate composites are commonly injection moulded and extruded, while plastics reinforced with long fibres are converted by hand lay-up or filament winding techniques.

TABLE 7.3
Plastics composites

Continuous phase	Disperse phase	
	Shape	Typical dispersed materials
Plastics and rubbers	Round	Air, gas, rubbers, plastics, glass beads
	Granular	Chalk, carbon black, wood flour, metal powders
	Plates	Glass, mica
	Fibres	Textile fibres, chopped filaments, woven cloth, non-woven cloth, glass fibres, carbon fibres

For certain physical properties, however, the volume fraction of the disperse phase rather than its shape or size is the principal controlling factor while in others there is a gradual transition in properties as we alter the disperse phase parameters. It is, therefore, sensible to consider both categories together while we are discussing their properties. On the other hand, when we come to the ways in which they are used it is more convenient to discuss them separately.

A convenient description of the disperse phase in particulate composites is 'fillers', and although the term often includes certain types of fibres and even paper, we shall restrict its use to particles which are: (i) roughly spherical; (ii) granular; (iii) flat platelets; and (iv) irregular particles whose L/D ratio does not exceed 10/1.

Generally speaking, the addition of powdered fillers to the polymer matrix results in an increase of modulus of rigidity accompanied by a lowering in tensile strength. The reason for this increase can be appreciated if we consider the equation developed by Einstein to explain the increase in viscosity observed in a liquid containing a dispersion of rigid particles in suspension. When the liquid is subjected to shear, a shearing force is produced across the diameter of the spheres causing them to rotate. The rotation absorbs energy and this effect shows up as an increase in viscosity.

The drop in tensile strength caused by the addition of fillers is thought generally to be due to the fact that, unlike the case of fibre reinforcement, the transfer of load from filler to matrix is relatively inefficient. Recently, however, materials have been produced in which both modulus and tensile strength are high. This is achieved by the use of fillers such as kaolin, calcium carbonate, asbestos, etc., which are coated with resin so that there is a gradient in modulus between that of the filler and that of the matrix.

Einstein's basic equation is expressed as follows:

$$\frac{\eta}{\eta_0} = 1 + \alpha V_f \tag{7.1}$$

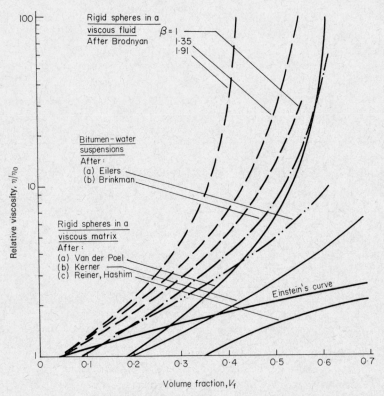

Fig. 7.2. Variation of relative viscosity with volume fraction of disperse phase.

where α is a constant (under these conditions $\alpha = 2\cdot5$), η_0 is the viscosity of the suspending matrix and V_f is the volume fraction of the dispersed spheres; this expression is independent of the size of the spheres. Einstein's equation is, however, limited to relatively low levels of V_f. Also, several additional factors are likely to increase the viscosity of the system including solvation and aggregation.

A number of modifications have been made to the basic equation in an attempt to extend its usefulness and also to take into account 'non-ideal' fillers. These include non-spherical rigid particles, viscous and elastic spheres in a viscous matrix, rigid spheres in viscous and elastic matrices and the extreme case of an elastic matrix containing spherical voids.

An accurate theoretical treatment of the behaviour of particulate composites is extremely difficult to achieve and many of the offerings have been highly empirical attempts to provide equations which will fit the available experimental data.

Some examples of the modified forms of equation 7.1 are as follows:

(i) Brinkman:
$$\frac{\eta}{\eta_0} = (1 - V_f)^{-5/2} \tag{7.2}$$

(ii) Eiler:
$$\frac{\eta}{\eta_0} = \left(1 + \frac{1 \cdot 25\, V_f}{1 - 1 \cdot 28\, V_f}\right)^2 \tag{7.3}$$

(iii) Mooney:
$$\frac{\eta}{\eta_0} = \frac{2 \cdot 5\, V_f}{1 - \beta V_f} \tag{7.4}$$

TABLE 7.4

Effect of variations in the disperse phase on the mechanical properties of polymeric composites

| Composite system | | | Disperse phase variables changing: | | |
| Phase | | | Shape—Plate to round to fibre | Size—Increasing | Concentration—Increasing |
Continuous	Disperse	Property			
Rubber	Carbon black	Tensile strength		Decrease	
		Shear modulus			Increase
		Tear strength		Decrease	Increase
Plastic	Elastomeric particles	Tensile strength		Decrease	Can increase or decrease
		Stiffness			Increase
		Impact strength			Increase
	Rigid particles	Tensile strength			Often decreases
		Stiffness Young's modulus	Decrease		Increase Increase
		Shear modulus	Decrease		

Eiler's equation reduces to equation 7.1 as $V_f \to 0$, and $\eta/\eta_0 \to \infty$, as $V_f \to 0.78$, which corresponds reasonably well with the fully close-packed state. In the case of Mooney's equation β is a hydrodynamic interaction factor with a value of 1 for 100% spheres increasing to 1·35 for close-packed spheres ($V_f = 0.74$) and to 1·91 for loosely packed spheres which

TABLE 7.5   **Properties**

Matrix material	C. black	PVC	Natural rubber	Nitrile rubber	SBR	Thermoset resins	Mineral fillers	Oil
						Additives		
Natural rubber	*					*		
						*		
						*	*	
				*				
Nitrile rubbers			*					
			*					
					*			
	*		*				*	
			*					
						*		
Polychloroprene	*					*		
						*		
						*	*	
S.B. rubbers	*					*		
						*		
						*		*
Polybutadiene rubbers	*					*		
Ethylene/propylene rubbers	*							
						*		
						*	*	

[a] Except for high levels of nitrile rubber.

are just free to move. As with equation 7.3, Mooney's expression reduces to equation 7.1 as $V_f \to 0$.

Modifications to eqn. 7.1 to allow its use with rubber systems reinforced with carbon black were developed by, among others, Vand, Guth, Smallwood and Simha. Smallwood's equation:

$$E^* = E(1 + 2.5\,V_f) \tag{7.5}$$

related Young's modulus E^* of the filled rubber at very small elongations and relatively low filler loadings to that of the unfilled rubber. The behaviour of materials with V_f up to 0·30 were described by a further modification, due to Guth, which takes into account higher volume fraction terms. Systems containing other fillers—mostly forms of carbon black—

...ber blends and mixtures

			Property change					
strength	*Abrasion resistance*	*Hardness*	*Chemical resistance*	*Aging resistance*	*Low temp. properties*	*Ozone resistance*	*Remarks*
		up	up	up			
vn		up	up	up			
		up					
						up	
vn	down	down	down[a]				[a]
		up	up				
			down	none	up		
ie	down	none	down				
ie	down	none	down				
up	up	up	up	up			See Fig. 3.2
		up	up				
up							
up							
		up					
up							} See Fig. 3.3
none							
		up					
		up					} See Fig. 3.4
			up				

were found to obey more closely an expression developed for rod-shaped particles.

$$E^* = E(1 + 0·67\Phi V_f + 1·62\Phi^2 V_f^2) \tag{7.6}$$

where Φ is a shape factor representing the ratio of length to diameter of asymmetric particles.

The way in which some of these expressions match up to the experimental data is shown in Fig. 7.2, together with the results of more elaborate treatments by Kerner, van de Poel, Reiner and Hashin.

These relatively simple treatments tend to indicate that the volume fraction of the disperse phase is the only factor which contributes to the flow characteristics of the composite. However, we know that shape and size are important controllers of the properties of metallic composites (cf. Section 1.3.3). It was also shown, when we considered mixing and compounding, that the degree of dispersion is an important factor. Tables 7.4 and 7.5 summarise the results of altering some of these parameters on the properties of polymeric composites, and Table 7.6 illustrates the effect of additives on polystyrene.

TABLE 7.6

The effect of additives on the physical properties of polystyrene

	Property				
Material	Elastic modulus ($kgf\,mm^{-2}$)	Tensile strength ($kgf\,mm^{-2}$)	Flexural strength ($kgf\,mm^{-2}$)	Impact strength (Izod) ($ft.\,lbf\,in^{-1}$)	Hardness (Rockwell)
Polystyrene (unmodified)	280–350	3·2–5·0	5–10·5	0·2–0·5	M65–80
Polystyrene (rubber modified high impact)	270–320	2·1–3·5	3·2–4·2	0·8–1·7	M40–60
Polystyrene/ copolymer blends (ABS)	190–270	3·9–5·2	5·7–8·0	2·5–7·5	R90–105

When the length of the fibre reinforcement exceeds about 10 times its diameter another factor becomes important—fibre orientation. The strength and stiffness of the composite in any one direction is proportional to the volume of the fibres aligned in that particular direction (Fig. 7.3). We may therefore define a quantity, the 'efficiency' of reinforcement η, which will have an effect on the ultimate properties of the material, and whose value depends on the orientation of the fibre and the direction of the applied stress. An illustration of this is given in Fig. 7.4. Here the contribution of V_f to the modulus E has been plotted for a series of composites: (a) polyester/carbon fibre; (b) polyester/glass fibre; (c) polycarbonate/glass fibre; (d) bitumen/silica. Superimposed on these experimental results are curves relating E and V_f at values of η corresponding to two extremes of unidirectional and random orientation. We see that in the case of the polyester/glass system, where the glass is present either as woven mat or chopped strands ($E_f = 0.3\text{--}0.4$) the experimental values do, in fact, fall between the bounds of $\eta = 0.2\text{--}1.0$. The polyester/carbon fibre composites contain fibres whose orientation is largely unidirectional ($\eta = 1$), and the

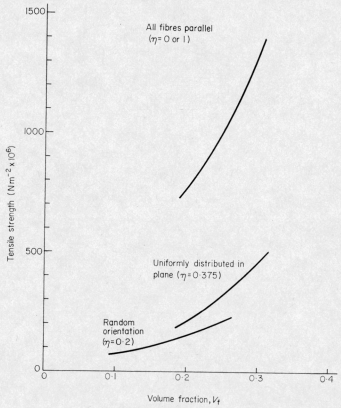

Fig. 7.3. Effect of glass content (V_f) and orientation on the tensile strength of glass/resin composites.

experimental values are found to lie close to the curve representing $\eta = 1$. The other two curves (c) and (d) contain respectively short fibres and 'granular' fillers.

7.2.3 Thermal properties

The thermal expansion behaviour of polymeric composites is important for two reasons. In the first place, since polymers have a relatively high coefficient of expansion, the inclusion of an 'inert' disperse phase provides us with a convenient means of reducing their high expansion; a factor which can be of considerable importance in certain precision engineering

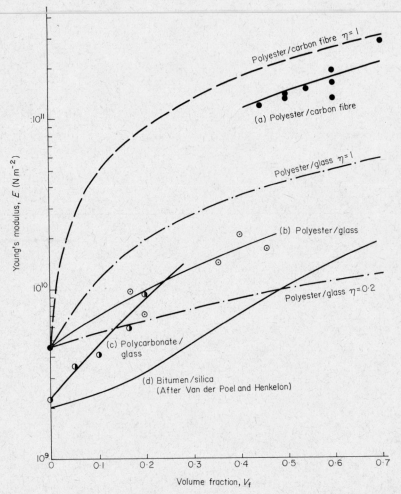

FIG. 7.4. Young's modulus of composites as a function of volume fraction of the disperse phase.

applications. In the second place, since there is an approximate relationship between elastic moduli and thermal expansivities, and where a relationship between bulk modulus and hardness also exists, measurements of the coefficients of expansion of materials which are relatively easy to carry out could provide us with information on such properties as stiffness, bulk modulus and hardness.

As we found with mechanical properties, many attempts have been made at developing acceptable equations to describe the thermal expansion of filled polymers, and the merits of a number of these have been discussed by Holliday and Robinson. A wealth of experimental data exists in the

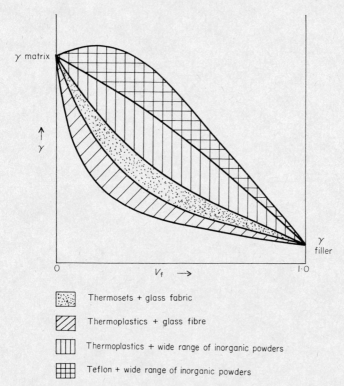

Thermosets + glass fabric

Thermoplastics + glass fibre

Thermoplastics + wide range of inorganic powders

Teflon + wide range of inorganic powders

FIG. 7.5. Coefficient of cubical expansion γ *vs* V_f for a range of materials. (Courtesy L. Holliday and J. D. Robinson.)

literature, and this has been collected and reduced to simplified block diagrams. Figure 7.5 shows the coefficient of cubical expansion (γ_c) for a variety of composites. The relationship between γ_c and bulk modulus (K) is shown in Fig. 7.6, and it can be seen that there is a reasonable agreement over a wide range of values for a considerable variety of materials.

As would be expected, the thermal conductivity of polymeric materials can be controlled by the addition of both fillers and fibres. The effect of the former is dependent mainly on the conductivity and the V_f of the disperse phase, while that of the latter is influenced additionally by the orientation of the fibres. Some values are given in Table 7.7.

7.2.4 Electrical properties

The remarks previously made about particulate composites refer in broad terms to their electrical properties. Generally (and particularly with the

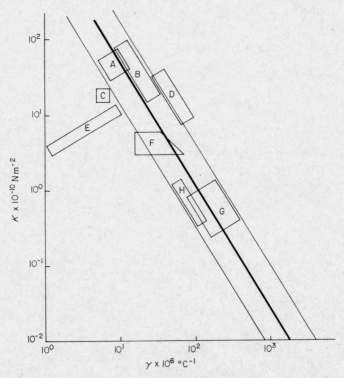

FIG. 7.6. Relationship between bulk modulus K and coefficient of cubical expansion (γ) for a range of materials. (Courtesy L. Holliday and J. D. Robinson.)

Key: A. Ceramics   E. Glasses
      B. Metallic elements F. Rare earths
      C. Silicon   G. Plastics
      D. Alloys   H. Plastic/glass composites

fibre reinforced materials), the electrical properties depend on the type of resin, the properties of the fibres, the quality of adhesion between the fibre and resin, humidity and fibre orientation.

TABLE 7.7

Thermal conductivity of polymeric composites

Composite	V_f	Thermal conductivity $(W\ m^{-1}\ °K^{-1})$	Direction of heat flow
Polyester resin	0	$20\text{–}25 \times 10^{-2}$	Any
Polyester/chopped glass fibre	17	$8\text{–}12 \times 10^{-3}$	Across fibres
Polyester/woven glass mat	46	$12\text{–}16 \times 10^{-3}$	Across fibres
Polyester/aligned carbon fibres	60	62^a	Across fibres
Polyester/aligned carbon fibres	60	1700^a	Along fibres
Epoxy resin	0	$15\text{–}130 \times 10^{-2}$	Any
Epoxy/glass fabric	—	40×10^{-2}	Across fibres
Epoxy/graphite fibres	—	60×10^{-2}	Across fibres
Epoxy/graphite fibres	—	16	Along fibres
Epoxy/PRD-49-111^b fabric	46	20×10^{-2}	Across fibres
Epoxy/PRD-49-111^b fabric	46	85×10^{-2}	Along fibres

^{a} Values quoted as $W\ m^{-2}\ °K^{-1}$.
^{b} Du Pont trade mark.

As a first approximation the dielectric properties are governed by an equation of the following type:

$$\log \varepsilon_c^* = V_f \log \varepsilon_f^* + (1 - V_f) \log \varepsilon_r^* \tag{7.7}$$

where ε_c^*, ε_f^* and ε_r^* refer to the dielectric constants of the composite, filler and resin respectively.

7.3 USING DISPERSION COMPOSITES

7.3.1 Introduction

Table 7.4 has shown us that the range of dispersion composites is enormous. The applications in which composites are successfully used is also enormous; they are particularly valuable in aerospace and the so-called 'engineering' plastics applications, and foamed materials especially are finding increasing favour in building, furniture, packaging and household applications. We shall, therefore, have to be content with a brief survey of some illustrative examples of three types of dispersion composites: (a) filled materials; (b) short and medium length fibre reinforced materials; (c) continuous filament composites.

7.3.2 Filled materials

Some examples of this type of material have already been discussed elsewhere in different contexts; the incorporation of additives and fillers

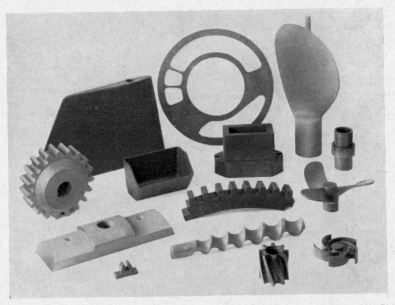

Fig. 7.7. Some examples of filled nylon components (Courtesy Polypenco Ltd., Welwyn Garden City.)

(Chapter 3), moulding of expanded materials (Chapter 4) and the extrusion of composite materials (Chapter 5). It is clear that, with the exception of the techniques developed for the production of certain foams and expanded materials, this category of composites can generally—with slight modifications—be processed by conventional fabrication techniques. Therefore, it is sufficient for us to deal with the characteristics of the material itself rather than the method of conversion.

For many engineering components ABS is quite adequate, and it has been used successfully for the manufacture of components in the automobile industry. However, there is a growing need for high specification materials for use in cars, including lamp housings and under-bonnet components. New grades of filled Nylon 11 containing up to 30% glass spheres and a mixture of 30% glass spheres with 20% glass fibres are proving more suitable than ABS in car light units where high temperatures can occur. These nylons also have good impact resistance and those containing only glass spheres are capable of being moulded with a high surface finish—suitable for electroplating. Also their 'self-extinguishing'

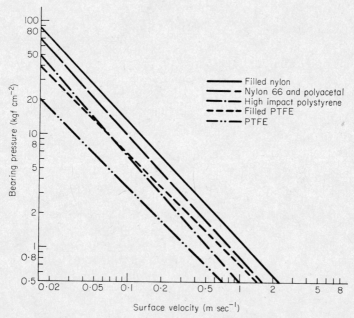

Fig. 7.8(a). Maximum allowable bearing pressure versus surface velocity. Operation is continuous and unlubricated at 23°C and 50% RH. (Courtesy Polypenco Ltd., Welwyn Garden City.)

rating to ASTM D635 makes them attractive for electrical and electronic components where the risk of fire must be minimised. Due to their good mechanical and surface characteristics glass spheres in resins are also used in dentistry. Colour matching is easily carried out by using spheres of different coloured glass.

Nylon and PTFE are increasingly becoming combined with such additives as molybdenum disulphide, graphite and mica to produce bearing materials with greatly improved dimensional stability and wearing characteristics. Although the nylon composites can be processed by injection moulding and extrusion, a powdered formulation is now available which, like the filled PTFE compounds, is converted into components by precision pressing and sintering techniques (cf. Section 6.3.2). Sintering without fillers can be used to produce components of controlled porosity which are capable of absorbing and retaining liquids. A typical application of such polymer/gas composites is in the production of 'self-inking' rollers for 'print-out' equipment. Figure 7.7 shows some examples of filled

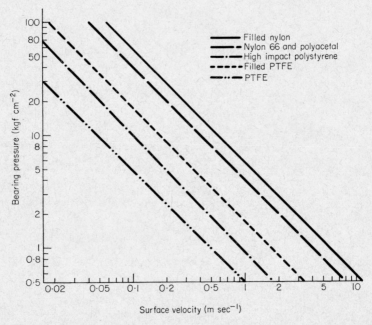

FIG. 7.8(b). Maximum allowable bearing pressure versus surface velocity. Operation is continuous; intermittent lubrication at 23°C and 50% RH. (Courtesy Polypenco Ltd., Welwyn Garden City.)

nylon components, and Fig. 7.8 illustrates the performance of a range of filled and unfilled bearing materials operating continuously.

7.3.3 Short and medium length fibre reinforced materials

A wide range of manufacturing methods is available for these materials, which are summarised in Table 7.8. Although, as has already been indicated (cf. Chapter 5 and Sections 7.2.2 and 7.3.2), chopped glass fibres, carbon fibres and various other synthetic high modulus/high strength fibres are widely used in thermoplastic polymers, reinforced thermosets are at

TABLE 7.8

Fabrication methods used with fibre-reinforced polymeric composites

Process	Advantages and disadvantages of the process														Applications								
	High capital cost	High labour cost	Fast production rate	Can mould undercuts	One good surface	Two good surfaces	Good detail pick-up	Thickness control	Wasteful in material	Much finishing reqd.	Easy to modify	Large mouldings	Low glass content	Can mould complex shapes	Car bodies	Boat hulls	Ducting	Pressure vessels	Roofing panels	Domes	Components	Sheet, rod stock	Any long runs
Wet lay-up	−	+	−	+	+			+			+	+	+	+	*	*							
Spraying	+		+		+			+	+		+	+	−	+			*						*
Filament winding	+	+	+					+	+			+						*					
Platen moulding	+	−	−			+		+				−							*				
Continuous moulding	−		+					−											*			*	
Injection moulding	+		+	+	+	+	+		−	−	−	−	+	+			*		*	*	*		*
Centrifugal laminating		+							−					−						*	*		*
Flexible bag moulding														−						*	*		
Matched-die moulding	+		+	+	+	+	+	+		−			−				*						*
Flexible plunger moulding	−			+	+	+	−	+				−	−										

Note: The use of + supports the statement; − does not, e.g. + denotes high capital cost, − denotes low capital cost, present the most extensively used fibre reinforced plastics. The most commonly used resins are polyesters and epoxies, and we shall concentrate on them for our examples.

Broadly speaking, we can divide the materials into three classes: (a) dough moulding compounds; (b) sheet moulding compounds; and (c) glass resin laminates formed by hand or wet lay-up techniques.

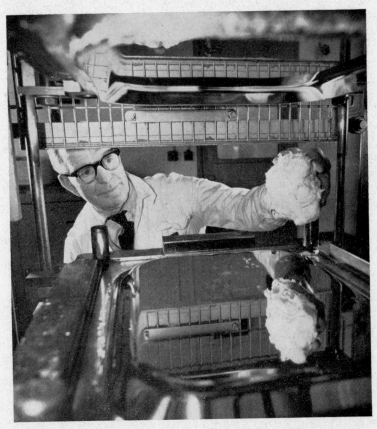

FIG. 7.9. Loading a moulding press with dough moulding compound. (Courtesy B.P. Chemicals International Ltd.)

(a)   Dough moulding compounds (DMC)

These materials are mixtures of polyester resin, chopped glass fibre, catalyst and fillers, although other fibrous reinforcements including, e.g. sisal, are sometimes used.

Mouldings from DMC can be produced which are completely free of moulding shrinkage or stresses, and which can perform satisfactorily over a range of temperatures extending from −60 to +150°C.

(b) Sheet moulding compounds (SMC)

The composition of these materials is similar to the DMC, but they are usually supplied as 'pre-pregs': these are chopped strand mats pre-impregnated with the resin compounds. Like the DMC, SMC are used in

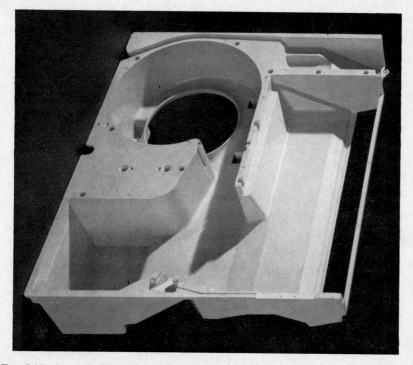

Fig. 7.10. A cooker hood chassis press moulded in one piece from dough moulding compound. (Courtesy B.P. Chemicals International Ltd.)

applications where the following properties are needed: good surface finish; low coefficient of thermal expansion; good temperature performance; excellent electrical properties; high strength and high stiffness; low pressure moulding and rapid cure.

Some examples of successful applications of DMC and SMC are shown in Figs. 7.9, 7.10, 7.11 and 7.12. The range includes precision engineering

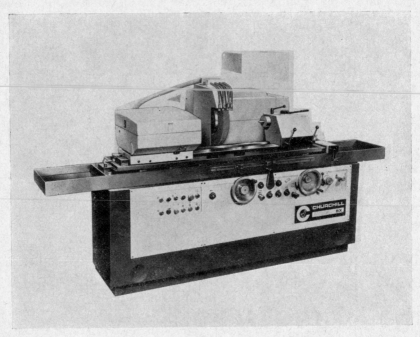

FIG. 7.11. The Churchill universal grinding machine. This machine incorporates a total of eighteen components which are moulded from DMC. (Courtesy B.P. Chemicals International Ltd.)

FIG. 7.12. Reinforced plastics dust covers moulded in polyester resin reinforced with glass fibre. (Courtesy B.P. Chemicals International Ltd.)

Fig. 7.13(a). H.M.S. *Wilton*, the Royal Navy's first plastics minehunter made from glass reinforced polyester. (Courtesy Ministry of Defence.)

Fig. 7.13(b). A view of the hull mould of H.M.S. *Wilton*. (Courtesy B.P. Chemicals International Ltd.)

components, automobile engine components and body parts, electrical components and domestic equipment of all kinds. Both DMC and SMC are formed by such techniques as compression moulding, ram injection moulding and transfer moulding, and can be successfully machined.

(c) Hand or wet lay-up techniques

One of the chief advantages offered to the designer by the use of glass/ resin laminates is the ability to design and fabricate large structures as a whole, rather than as an assembly of components. This advantage is

FIG. 7.13(c). Work on a deck section of H.M.S. *Wilton.* (Courtesy B.P. Chemicals International Ltd.)

illustrated particularly well in marine applications, and Fig. 7.13(a) shows a general view of H.M.S. *Wilton*, the Navy's first plastics minehunter. An interior view of the hull, which is 50 metres in length (Fig. 7.13(b)), illustrates the way in which the material thickness at different locations can be varied, and extra strength built in at any point and in any direction. Figure 7.13(c) which gives a detail of men working on the hull, shows how extra rigidity is achieved by the integral moulding of ribs into the laminate. These ribs can be solid or hollow; for solid ribs a core of glass fibre, wood or foamed plastic, is used, while hollow ribs are made by the use of re-movable flexible tube, or a simple former made from cardboard or other inexpensive material.

Figure 7.13(d) gives a view of another marine application—the outer structures of an 'Oberon' class submarine.

Figure 7.13(c) incidentally also illustrates the most commonly used method of producing GRP (glass reinforced plastics) structures. The mould surface—often of wood—is coated with a release agent. The laminate consists firstly of a gel coat applied to the mould surface: this is to ensure a smooth surface free from protruding fibre-ends. After this the moulding is built up by applying successive layers of resin and glass mat,

FIG. 7.13(d). Construction of the outer structures of an 'Oberon' class submarine in the Royal Naval dockyard. (Courtesy B.P. Chemicals International Ltd.)

a specially profiled roller being used to remove the entrapped air and ensure thorough wetting of the glass fibres. An alternative method of fabrication is spraying (Table 7.6) in which the resin and hardener, and the chopped glass fibre, are sprayed simultaneously onto the mould. The technique is illustrated in Fig. 7.14 which shows the manufacture of a reinforced plastics silo.

Because glass fibre is relatively cheap it is economical to distribute it evenly throughout the laminate. With carbon fibres, however, their high rigidity and high cost make it convenient to achieve reinforcement by incorporating strips of carbon fibres at strategic points in the structure (Fig. 7.15).

7.3.4 Continuous filament composites

Three-dimensional shapes can be made, using glass filaments or carbon fibres, by passing rovings through a bath of activated resin and then winding the impregnated rovings onto a rotating mandrel. This method is

FIG. 7.14. The spraying technique used in the manufacture of glass reinforced polyester silos. (Courtesy B.P. Chemicals International Ltd.)

particularly suited to making symmetrical shapes, and has been used to produce cylindrical tanks, high pressure vessels, as well as submarine and rocket motor casings (Fig. 7.16(a)). Asymmetric articles have also been

FIG. 7.15. The use of carbon fibres in reinforced plastics. (*right*) *Daffodil II*, the capsule of the high altitude hot air balloon which smashed the world altitude record in January 1974. Carbon fibres in high modulus ribbon form were applied, in a grid pattern, around the exterior of the capsule to provide additional light-weight stiffness and strength. (*top*) A typical G.R.P. panel, stiffened with two tows of carbon fibre in a knitted glass fibre tape, showing the depth and uniformity of the carbon fibre rib and the additional reinforcement provided by the glass at cross-over points. The insert is a close-up of the continuous carbon fibre tows in a knitted glass fibre tape. (Courtesy Courtaulds Ltd., London.)

FIG. 7.16(a). Filament wound GRP wire storage tank. Capacity 9100 litres. (Courtesy Scott Bader Co. Ltd., Wellingborough.)

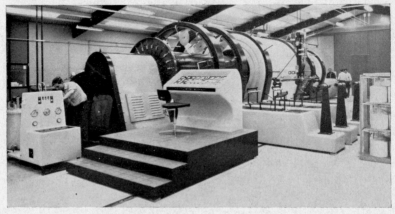

FIG 7.16(b). Continuous filament winding machine for circular cross-section. Capacity 14–28 m per hour. (Courtesy Scott Bader Co. Ltd., Wellingborough.)

produced in this way, and sections of aircraft wings made by filament winding are currently in service in the U.S.

Filament winding is also used, as an alternative to rotational casting, to produce long continuous lengths of pipes in diameters ranging from a few centimetres to several metres (Fig. 7.16(b)).

7.4 LAYER STRUCTURES

The main reasons for making layer structures include the following:

(a) improvement of mechanical properties;
(b) enhancement of barrier characteristics;
(c) promotion of adhesion;
(d) as a means of protection—either for a delicate surface or for printing;
(e) the achievement of a decorative effect.

As a rule quite a number of these benefits are realised when a layer structure is created, although occasionally one of them will predominate to a very marked extent. For example, in a laminate based on aluminium foil, the most significant performance factor is impermeability—to radiation as well as to gases and to vapours. The secondary but nevertheless important benefits are heat-sealability and the 'metallic' decorative effect. Many other examples are possible and it is clear that this variety of layer structures is potentially very large indeed. In fact, a glance through the recent technical literature seems to suggest that some benefits result—or at least are claimed—from sticking almost every imaginable material to another.

In order, however, to cover the ground as adequately as space will allow, it will be sufficient to split the field into a number of relatively arbitrary categories and consider some examples of both typical and novel materials and applications in each of them. A simple yet realistic method of achieving this subdivision is to use a shape and complexity factor. This allows the field to be split into three:

(a) flat sheets—including coated film and laminates;
(b) simple mouldings—either produced in one operation or built up in a series of steps;
(c) structures—buildings, vehicles, furniture, etc.

7.4.1 Flat sheets

A large proportion of the layer structures made either by coating or combining sheets or films of different materials were created for the packaging industry, and thus their function in the simplest possible terms is to separate the product from the environment. The performance requirements of the laminate will depend on what is to be packed and the degree of protection needed, and these are summarised briefly in Table 8.2. There are, of course, secondary—but often extremely important—characteristics which the packaging must have. An obvious one is that it must sell the product, and therefore have a pleasing and attractive appearance which must not deteriorate. Also it must fall within the right price range. Less obvious, perhaps, but equally important, is that the laminate should work properly on conventional packaging machinery. To achieve this an extra ply is sometimes added. Table 7.9 gives a brief selection of some material combinations used in packaging. It is obvious even from this small selection that the choice of the most suitable combination can be a major problem since, as we have seen, the considerations include not only mechanical properties, handling and visual characteristics, but also cost as well as barrier properties.

TABLE 7.9

A selection of flexible laminates used in packaging

Structure	Characteristics
(1) Simple laminates	
Paper/plastic (i.e. polyethylene, polypropylene, polyvinylidene chloride, ethylene vinyl acetate, etc.)	Opaque
Cellulose/plastic	Transparent, good barrier
Aluminium foil/plastic	Opaque, impermeable
(2) Multiple laminates	
Ionomer/nylon	Transparent
Ionomer/polyester	strong,
Ionomer/fluorocarbon	good barrier
Polypropylene/copolymer/homopolymer/ copolymer	Coextruded: transparent, good heat seal, overwrapping
Polypropylene/copolymer/homopolymer copolymer/cellophane	Good for gussetted seals
Oriented polypropylene/polyvinylidene chloride, polyethylene, polyvinyl acetate, etc.	Transparent, good heat seal
High density polyethylene/foil/polyester	High strength, opaque, impermeable
Cellulosic/board	High gloss, opaque

Even when attempting to design a material with specific barrier characteristics one has the choice of fulfilling the requirements in several ways. These are dealt with in greater detail in Section 8.3 but in almost all cases a compromise is required between such factors as: (a) cost; (b) mechanical strength; (c) performance on machinery; and (d) permeability.

The methods of making layer structures are by now quite familiar, and it is sufficient to remind ourselves that they can be made by one or more techniques based on the following processes: (i) extrusion; (ii) calendering; (iii) melt roll coating; (iv) dip coating; (v) adhesive coating and laminating; (vi) heat-laminating.

Although we have been considering these materials primarily in the context of packaging, many of them are widely used in other fields. For instance, metallised polyester films are used in automobiles as decorative trim, and in meteorological balloons and other aerospace applications where strength and lightness are important. When combined with various substrates and coated with coloured lacquers these laminates provide a range of decorative materials which are used for clothing and many other domestic applications. Normally the metallic coating is aluminium applied by vacuum metallisation, but more recently a process has been developed for the continuous deposition of silver on to a cellulose acetate base which is then reverse-coated with an adhesive.

So much for flat-sheet materials, but we must remember also that some of the three-dimensional objects we shall be considering in later sections can be formed from laminated sheets, either by thermoforming or by stamping techniques.

7.4.2 Simple mouldings

Metal stamping techniques have been applied to a range of plastics materials (cf. Section 6.3.3), and, provided the various components of the laminate are capable of extension, there is no reason why the process should not be used for laminates. Cold-formed vessels have been formed from laminates of reinforced thermoset resin cores and face sheets consisting of an inner layer of metal foil and an outer layer of thermoplastic resin. This example will be discussed in some detail as it provides an interesting example of the interplay between the design of the finished article, the material and the forming process.

During the cold forming operation it is necessary for the core to be in an uncured state, since it must not only flow but also be subjected to a complex strain history. Equally the core alone cannot be deep drawn because the uncured thermosetting resin has a low tensile strength, and so we need the face plies. These support the tensile stresses in the core, and therefore the controlling parameter becomes the hold-down pressure necessary to avoid wrinkling of the face sheets. Since the laminate is open at its periphery, the core must withstand the hold-down pressure without any appreciable squeezing out.

The laminate consisted of a sandwich of PVC/aluminium foil on either side of a reinforced epoxy resin core, and in its final cured state had a rigidity compared with that of 10^{-6} m steel. Table 7.10 shows a comparison of the forces necessary for drawing the materials separately and in combination.

TABLE 7.10

Cold forming forces required for laminate, plastics and metal sheet

(After H. J. Oswald, P. Kosh, H. L. Li and D. C. Prevorsek.)

Material	Thickness $(10^{-5}\ m)$	Hold down force $(10^4\ N\ m^{-2})$	Punch force $(10^4\ N\ m^{-2})$
PVC	5	245	126
Al sheet	6·25	350	154
C steel sheet	10	539	770
Plastics laminate	25·1	259	147

Most of the mechanical properties of the laminate were also found to benefit by curing the material under external pressure. The most significant improvements were in tensile strength and modulus, which increased by 40% and 18% respectively.

Some additional improvement to impart strength resulted from the use of a flexible epoxy resin core, although at the expense of some rigidity.

Boxes and other articles are increasingly made from cold-rolled steel laminated to PVC film. These products are highly decorative and also possess good chemical resistance. As we saw in Section 4.2.4 it is possible to coat fabricated metal articles with a thermoplastic resin and so obtain a layer structure by yet another technique.

Now we must consider briefly the very large field of applications in which an article or container may be made by one of the following operations:

(a) formed from a combination of materials in one operation;
(b) formed and then coated;
(c) formed and then combined with other materials or even with other formed components.

In category (a) bottles have been made by the simultaneous coaxial extrusion of a parison composed of two different materials, i.e. an outer layer of nylon on polyethylene. It has also been claimed that a similar effect can be achieved by rotational casting, where two powdered materials of different densities are melted together, the heavier diffusing through the melt to form the outer skin.

Containers for instruments, or other applications where detailed contouring of the inside of the container is important without affecting the outside appearance, have been produced from high density polyethylene using a double wall blow moulding technique. The double wall contains enclosed air which provides cushioning during the forming process, and also makes the finished package more robust.

As in the case of sheet materials, the properties of moulded containers can be adjusted by coating (b). This is discussed from the point of view of barrier improvement in Chapter 8. The added layer can, of course, also provide the additional effects of decoration and protection to a sensitive or printed surface. Experience has shown, however, that with relatively brittle materials, the impact resistance can decrease on coating; it appears that the coating acts as a stress raiser and prevents the rapid dissipation of the energy of impact.

It is reasonable to suggest that plastics articles coated with vacuum- or electro-deposited metal are layer structures, partly because the whole behaves differently from the individual components, and also because there is usually an interaction between the manufacturing conditions under which the separate components are made, and the performance of the composite.

An important factor in the production of metallised components is the state of the interface between the plastic and the metal (cf. Section 7.3.2). In particular the moulding parameters have a considerable effect on the adhesion of the metal layer (see Table 7.11).

The automobile industry provides one of the more fruitful fields for the use of moulded, coated components. Figure 7.17, which compares vehicle rear light assemblies made from a variety of materials, also represents an example of the type of structure mentioned in category (c).

TABLE 7.11

Effect of moulding conditions on the adhesion of electro-plated metals to ABS plastics

Factor	Effect on adhesion
Melt temperature	Increases with melt temperature becoming more marked as rate of cavity fill becomes slower
Injection pressure	Lower pressure gives better adhesion
Wall thickness	Improves with thickness

FIG. 7.17. A comparison of vehicle rear light assemblies. (*left*) Made from steel and glass. (*right*) Metallised ABS and moulded acrylic. (Courtesy Marbon Europe N.V., Borg-Warner Chemicals, Amsterdam.)

7.4.3 Structures

In this area the principal criteria are lightness and strength, and it is here that various combinations of plastics and of plastics with other materials

FIG. 7.18(a). Interior view of an insulated GRP container. (Courtesy B.P. Chemicals International Ltd.)

FIG. 7.18(b). Insulated containers manufactured with GRP panels. (Courtesy B.P. Chemicals International Ltd.)

are making the most striking impact. A selection of examples is given in Figs. 7.18 and 7.19, where the various combinations of plastics and other materials are used to provide decorative as well as structural effects.

Figure 7.18 shows two views of a freight-liner container in which the properties of strength, lightness and insulation are achieved by a layer

FIG. 7.19. The roof of the new Covent garden market. Made from 924 individual GRP/foamed core/GRP sandwich mouldings, the roof is the biggest GRP roof in Europe. (Courtesy Scott Bader Co. Ltd., Wellingborough.)

structure incorporating metal, wood, glass, air and plastics. Figure 7.19 shows a large structure made from laminated plastics.

Finally, of course, there is the very large area of furniture and furnishings in which highly complex layer structures are built up using wood, metal, reinforced plastics, rigid and flexible foams, plastics films and textiles.

In this chapter it has only been possible to give the briefest coverage of an enormous field—materials in combination. However, it is in this field that the most interesting new engineering materials are likely to be created. Not only will it be possible to extend the range of attainable physical properties, but by a judicious choice of recipe a balance can be struck between the use of materials which are plentiful and those which are becoming scarce.

FURTHER READING

Carbon Fibres: their Composites and Applications, Plastics and Polymers Conference, Supplement No. 5, 1971.

Holister, G. S., and Thomas, C. (1966). *Fibre Reinforced Materials*, Applied Science, London.

Holliday, L. (Ed.) (1966). *Composite Materials*, Elsevier, Amsterdam.

Holliday, L., and Robinson, J. D. (1970), 'The Thermal Expansion of Composites Based on Polymers', *J. Mater. Sci.*, **8**, 301.

Oswald, H. J., Kosh, P., Li, H. L., and Prevorsek, D. C. (1970). *Modern Plastics*, March, 82.

N.P.L. Conference (1971). *The Properties of Fibre Composites*, I.P.C. Science & Technology Press, London.

Wake, W. C. (Ed.) (1971). *Fillers for Plastics*. Iliffe, London.

CHAPTER 8

DESIGNING WITH PLASTICS

8.1 INTRODUCTION

So far we have considered the nature of polymeric materials and the factors
which control their properties. We have also dealt with the various ways
in which they may be converted into useful articles. The successful application of plastics to the end product can be a complex and difficult exercise
with a dauntingly large array of variables. However, these can be sub-
stantially reduced and the problems of choice made much easier provided
certain requirements are clearly understood. The first is that in replacing a
familiar material by plastics the whole concept of the application must be
considered, since differences in material properties and manufacturing
methods will often have far reaching consequences in product design.

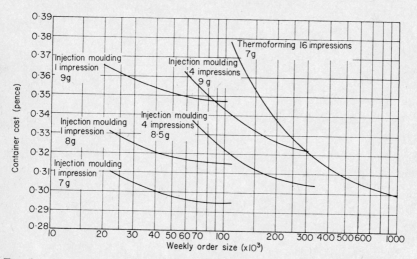

FIG. 8.1. Comparison between injection moulding and thermoforming a 175 ml
yogurt container. (Courtesy D. H. Latham, Shell Research Ltd.)

One rather trivial example is the use of plastics in beer crates. No one, in designing the new crate, remembered that barmaids often stand on up-turned crates to reach the top shelves. In the days of stiletto heels this proved almost disastrous!

In designing gear wheels, for instance, nylon seems to be an unlikely candidate due to its relatively low strength and rigidity compared with metals. With metal gears, however, the area of contact between meshing teeth is so small that extremely high pressures are developed. Nylon's greater flexibility allows the teeth to distort thus increasing the contact area and reducing the loading on each tooth. The lesson to be learned is that requirements should not be interpreted in terms of any particular material: it is the function that is important.

TABLE 8.1

Comparison of fabrication methods (the number required represents the accepted minimum economic output). (Courtesy P. C. Powell, I.C.I. Ltd., Plastics Division, Welwyn Garden City, Herts.)

Production method	Number required
Machining from stock	1–100
Sheet shaping	100–1 000
Rotational moulding	100–1 000
Extrusion[a]	300–3 000
Blow-moulding	1 000–10 000
Injection moulding	10 000–100 000

[a] Minimum length 300–3000 m.

The market for the product is also important, since the quantity of articles to be produced is one of the principal factors in determining the method of manufacture (Fig. 8.1 and Table 8.1). This factor is also important in product design since certain shapes are difficult to make by a particular process: injection moulding is unsuitable for the production of narrow-necked containers. Also, in multi-component assemblies, using perhaps a variety of different plastics, it is necessary to consider whether in the case of failure in service it is more convenient (cheaper in material and labour costs) to repair or replace some or all of the components.

Ideally, the problem should be handled by four people; the customer, the designer, the processer and the materials supplier. Usually several of these functions are combined or delegated and the relationship becomes one between the customer and the fabricator.

We have already stated that a number of variables are involved and it is helpful to deal with them under three broad headings: the polymer, the process, and the product.

8.2   THE MATERIAL: CHOOSING THE RIGHT ONE

Not all applications call for the ultimate in properties, but all applications call for a compromise, either on technical or economic grounds or both. It is important, therefore, to be quite precise about the requirements of the product so that a clear distinction may be made between the properties that are essential and those that are merely desirable. Some indication of these requirements can be obtained from a simple classification of the type shown in Table 8.2. Here the applications, all concerned with packaging, have been divided into four categories in ascending order of severity.

TABLE 8.2

Package specifications for different classes of chemicals

Category	Typical examples	Requirements
1	Inert powders, e.g. magnesium oxide	Simple mechanical barrier between product and environment
2	Volatile liquids, 'perfumed' solids or semi-solids, e.g. solvents or toothpaste	As for 1 plus barrier properties to prevent escape
3	Materials sensitive to environment, e.g. salt, foodstuffs, motor oil, household bleach	As for 1 plus barrier properties to prevent attack
4	Aggressive materials, e.g. caustic soda, detergents	As for 1 plus possibly 2 and 3. Also need product resistance

Alternatively, we could use the classification given in Table 8.3 where the division falls into four fields of activity: structures; components; protection and electrical.

Whichever method is used—and there are many others—a classification of this sort serves two important functions: using it makes one think more carefully about the product and also makes one realise how often a product's requirements can be satisfied by one of the relatively cheap large tonnage materials like polyethylene, polypropylene or PVC.

TABLE 8.3

Requirements of engineering materials

Application	Requirements
1. Structures	Generally speaking these will be load bearing to a greater or lesser degree
2. Components	(a) Non- or light-stress bearing components, i.e. structural parts; lids, facings, ducts, containers, casings, etc., as employed for instance in the field of domestic appliances; motor vehicles, aircraft, building, lighting and marine applications; small parts; plugs, knobs, vents, keys, seals, etc.
	(b) Stressed parts; i.e. gear wheels, cams, pressure vessels, impellers and fans, fasteners, lock parts, pulleys, hinges, bearings, etc.
3. Protection	Where chemical corrosion or thermal resistance is important, including water resistance
4. Electrical	Where, in addition to electrical insulation, mechanical functions of mechanism or mounting are also required

8.3 GENERAL MATERIAL CHARACTERISTICS

The selection of materials involves a number of stages in which their characteristics are examined more minutely as the field narrows. As a preliminary exercise, three main criteria provide a useful screening procedure. These are:

(a) whether or not optical transparency is needed;
(b) the resistance of the material in service (i.e. to chemical, thermal and other environmental damage);
(c) its mechanical properties.

Table 8.4 gives a survey of a selection of the commonly used plastics with the properties ranked for each material. 1 counts high, and therefore signifies, for example, the stiffest material or that with the best barrier performance (least permeability). Precise data are difficult to obtain because of differences both in grades of commercially available materials and in the methods of testing. Also, for reasons that will be discussed later, simple measurements of mechanical properties of polymers are only of very limited value. Where there is a considerable spread of results or overlap between materials in the Table they have been allocated the same number.

TABLE 8.4

Properties of plastics materials

Material	Density	Stiffness	Impact strength	Heat resistance	Clarity	Price	Barrier to: O$_2$	H$_2$O
				Property assessment (1 ranks high)				
TPX	17	8	4	2	2	4	11	7
Polypropylene	16	9	3	2	4	10	9	2
Polyethylene LD	15	12	1	9	4	13	10	3
Polyethylene MD	14	11	2	5	5	12	9	2
Surlyn A	13	13	1	8	3	5	9	3
EVA copolymers	12	13	1	7	3	8	12	10
Polyethylene HD	11	10	2	4	5	11	8	1
ABS	10	7	2	8	4	7	7	2
Nylon 6	10	8	2	6	3	2	1	6
Polystyrene	9	1	5	10	1	13	8	7
Polyphenylene oxide	8	4	3	1	6	1	4	8
SAN	7	1	4	9	2	9	6	—
XT polymers	6	5	4	—	4	3	5	6
Phenoxy	5	2	1	10	2	2	2	5
Polycarbonate	4	6	1	3	2	4	8	9
Cellulose acetate	3	7	5	9	1	6	4	10
Rigid PVC	2	3	4	11	1	10	3	4
Polyacetal	1	1	1	8	6	4	5	5

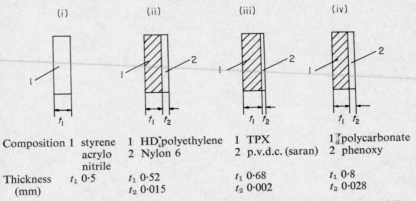

Composition 1 styrene   1 HD,polyethylene   1 TPX   1 ¼ polycarbonate
            acrylo   2 Nylon 6   2 p.v.d.c. (saran)   2 phenoxy
            nitrile

Thickness t_1 0·5 t_1 0·52 t_1 0·68 t_1 0·8
(mm) t_2 0·015 t_2 0·002 t_2 0·028

FIG. 8.2. A comparison between structures all having the same oxygen permeability,
i.e. 3·13 cm³/100 in²/mil/24 hr/atm.

Looking at the Table several things become apparent. In the first place, because of the variety of combinations of properties, the need to be precise in specifying requirements emerges again very strongly. Secondly, it is clear that there will be several ways of fulfilling the need. For example, in

TABLE 8.5

Comparisons between containers made from different materials

	Material				
Property	Poly-ethylene LD	Poly-styrene	Poly-carbonate	Poly-ethylene HD	Rigid PVC
Material					
O_2 permeability (cm³/100 in²/ mil/24 hr/atm	370	270	200	130	8
Thickness (mm) for equivalent O_2 permeability	15·4	10·9	8·1	5·3	0·38
Impact strength	high	poor	high	good	fair
Stiffness (E) N m⁻² × 10⁹	0·240	3·1	2·2	0·90	2·58
Price per kg	21·3	13·8	39·9	21·1	24·6
Density g/ml	0·92	1·05	1·2	0·96	1·4
Containers					
Containers with equivalent O_2 barrier characteristics					
(a) weight of container (kg)	11·2	9·0	7·7	4·0	0·42
(b) cost per container (£)	2·38	1·25	3·07	0·85	0·13

packaging, whether for foodstuffs, chemicals or medical applications, it may be necessary to contain volatile or oxidisable materials and prevent loss of components or attack from outside.

The appropriate barrier characteristics can be achieved in several ways. Firstly, we could contemplate a relatively thick-walled container made from a material which, although a comparatively poor barrier, is mechanically strong, cheap and easily processed on normally available equipment. Alternatively we could use a thinner section of a high barrier material and compromise on mechanical properties and cost. Again there is the possibility that the necessary characteristics are best realised by combining two or more materials. For example, packets for freeze-dried foods often are made from a multiple laminate including aluminium foil, paper and a heat-seal coating. Figure 8.2 illustrates some of the ways in which a chosen barrier performance may be achieved, and Table 8.5 presents a range of arbitrary containers. Some of these are clearly impossible to use either on the grounds of high cost or mechanical weakness. but are nevertheless included to illustrate the place of these factors in materials selection.

The selection procedure we have so far been operating is perfectly adequate for providing the potential plastics-user with information on the type of material which might be suitable for the application in mind. However, to be of real value this information needs to be supplemented by data from tests more closely simulating operational conditions.

8.4 PLASTICS DESIGN DATA

Thermoplastics are being increasingly used in applications involving exacting cost/performance requirements, such as bottle crates and containers, pipes and large liquid storage tanks. In designing articles of this kind certain types of questions need to be answered. *Will the article stand up to its service requirements, and what is the amount of distortion we can expect or tolerate? What is the minimum amount of material that can be used and yet expect the article to perform adequately?* Field experience is of value here, but in order to be able to design the article correctly with adequate safety factors, we need more reliable data on the properties of the material.

Metals—whose properties are familiar to most designers—are often treated as linear elastic solids, and are therefore characterised by only two constants: Young's modulus (E) and Poisson's ratio (v). Many of the standard test methods for plastics were originally developed for quality control purposes by the materials suppliers, who found it convenient to use the tests for such properties as strength, rigidity, hardness, wear, etc., which had been successfully developed with metals.

When a material, such as unplasticised PVC, is subjected to a tensile test and its performance compared with steel, its stress/strain characteristics are seen to be different in three important respects: (Fig. 8.3)

(a) the yield stress is lower;
(b) the yield strain is higher;
(c) below the yield point there is no region of constant slope.

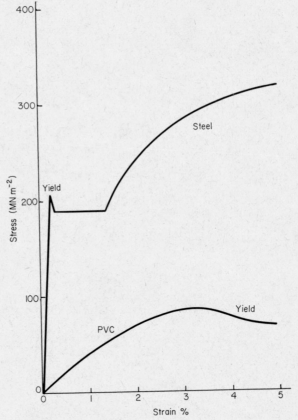

FIG. 8.3. Conventional stress/strain diagrams for steel and for unplasticised PVC. (Courtesy I.C.I. Ltd.)

We know from Chapter 2 that most plastics behave as non-linear viscoelastic solids, with the results that, for example

(i) properties such as stiffness will vary with time stress and temperature. Hence, if the test measurements illustrated in Fig. 8.3 were carried out at a higher temperature or a lower elongation rate the value of the modulus would decrease;

(ii) mechanical properties, including toughness, will be influenced by design and size of the component, design of the mould (Chapter 4), processing conditions (Chapters 4, 5 and 6) and service temperature;

(iii) the basic physical properties are affected by environmental effects such as thermal and oxidative aging, u.v. aging and chemical attack;

(iv) a change in a specific polymer parameter may affect both processability and basic physical properties, and both of these can interact in governing the behaviour of the fabricated product.

Having established that the behaviour of plastics depends on their previous history and the way in which the stress is applied, it is necessary either to devise new tests which have a sound theoretical backing, or attempts must be made to broaden the currently available standard tests to include changes in the various fundamental quantities, such as stress, strain, time and temperature.

The need for this type of approach was recognised by, among other organisations, the British Standards Institution which set up in 1965 a Technical Committee, PLC/36. The Committee's terms of reference were to consider the problem of selecting and presenting design data on plastics materials. In October 1967 a 'Draft British Standard Guide for Plastics Design Data' was circulated for comment throughout industry, and in 1970 the first seven parts of BS.4618:1970 'Recommendations for the presentation of plastics design data' were issued. The work is continuing, and more meaningful data are gradually becoming available. There are a number of ways in which we could consider the behaviour of polymers, but for our purposes it is convenient to divide the test conditions into two main areas:

(a) prolonged continuous stress, i.e. pipes, crates, tanks, etc.;
(b) short term stresses, i.e. yielding, brittle fracture or fatigue.

We shall treat them separately although it must be realised that in service they can be, and often are, superimposed.

8.5 MATERIALS UNDER PROLONGED STRESSES

Under conditions of constant stress, materials will, at appropriate levels of stress and temperature, exhibit a continuous deformation with increasing time. An example is the compression occurring in the bottom crates of a stack or the loading experienced at the base of a storage tank. This behaviour is called 'creep' and a typical creep curve is shown in Fig. 8.4.

With thermoplastics the secondary stage is often only a point of inflexion and the final tertiary stage is usually accompanied by crazing or cracking,

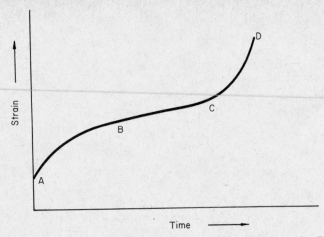

FIG. 8.4. Typical creep curve. Generally split up into three parts: A—B primary creep rate decreases linearly with time. B—C secondary stage where change in dimensions with time is constant; i.e. constant creep rate. C—D tertiary stage; creep rate increases until rupture occurs.

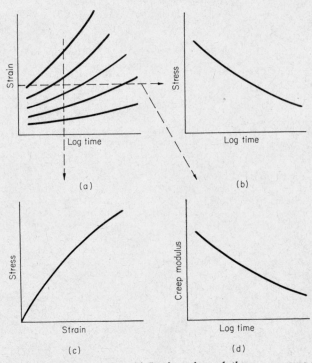

(a)

(b)

(c)

(d)

FIG. 8.5. Presentation of creep data. (a) Sections through the creep curves at constant time and constant strain give curves of (b) isometric stress/log time, (c) isochronous stress/strain and (d) creep modulus/log time. (Courtesy I.C.I. Ltd.)

or at higher stresses by the marked local reduction in cross-sectional area which is called 'necking'. Curves of the type shown in Fig. 8.4 are not normally used however since, although they illustrate creep behaviour correctly, the information provided is of limited value. Generally a series of creep curves is obtained, each at a series of test temperatures. Strain is

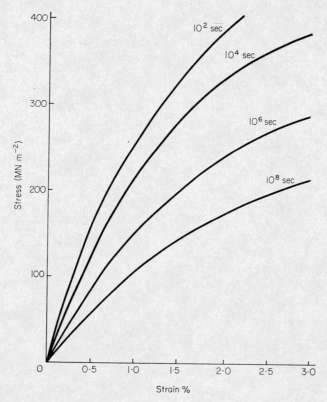

FIG. 8.6(a). Isochronous stress/strain curves for acetal copolymer at 20°C and 65% RH. (Courtesy I.C.I. Ltd.)

normally plotted as a linear scale and, because of the long test times involved, time is shown as a logarithmic scale (Fig. 8.5(a)).

The more rigid plastics such as polymethyl methacrylate have a relationship between strain, stress and time which can be represented by:

$$\varepsilon = \sigma f(t) \tag{8.1}$$

which implies linear viscoelastic behaviour.

However, as we have already seen, many plastics do not behave in this way; for instance, polypropylene shows non-linear viscoelasticity in most applications of practical importance, and its behaviour more closely approximates to:

$$\varepsilon = f(\sigma, t) \tag{8.2}$$

The creep data so far discussed have been obtained from samples tested under uniaxial tension, since (a) this represents the most convenient

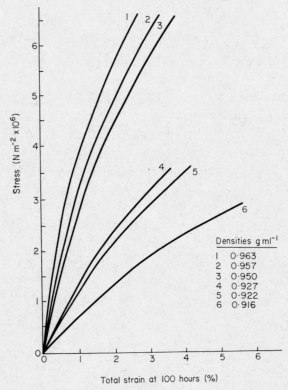

FIG. 8.6(b). Isochronous stress/strain curves for polyethylenes at 23°C.

method of testing, and (b) the results are in a form acceptable to the designer. Depending on the particular applications and materials involved it is sometimes preferable to present the data in another form to take account of particular aspects of the design problem.

Three such presentations of the basic creep data are shown in Figs. 8.5(b), (c) and (d).

8.5.1 Isochronous stress/strain curves

Cross plots at constant time (Fig. 8.5(c)) give isochronous stress/strain curves of the form shown in Fig. 8.6(a). An example when this type of presentation is useful is in designing a milk crate. We could reasonably

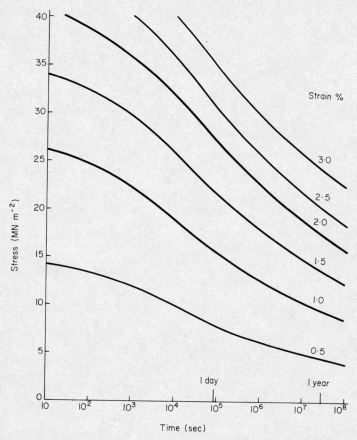

FIG. 8.7. Isometric stress/log time curves for acetal copolymer at 20°C and 65% RH. (Courtesy I.C.I. Ltd.)

assume that the moulding would not have to withstand continuous loading for more than 100 hr, and that after 100 hr the strain must not exceed 1%. The stress which can be sustained by each material without exceeding a strain limit of 1% can then be read from the curves in Fig. 8.6(b).

8.5.2   Isometric stress/time curves

A characteristic of thermoplastics is that above a certain stress further increases in stress produce a disproportionate increase in strain. This makes it important that we know which part of the performance curves cover the intended application. The use of isometric stress/time curves (Fig. 8.5(b)) makes it possible to read off easily the permissible stress which, if applied for a specified time, will not cause the specified maximum strain to be

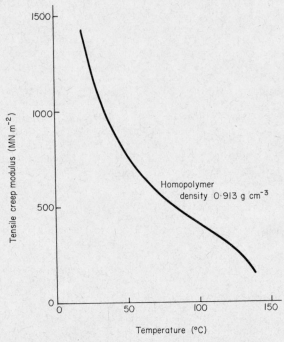

FIG. 8.8. 100—second creep modulus (0·2% strain)/temperature curve for propylene homopolymer, density 0·912 g cm⁻³. (Courtesy I.C.I. Ltd.)

exceeded. Such curves are constructed by plotting the points of intersection at constant strain on the family of creep curves shown in Fig. 8.5(a). Figure 8.7 shows a set of such curves for acetal copolymer.

8.5.3   The influence of temperature

The behaviour of plastics varies considerably with temperature, and a complete expression of the inter-relationships between stress, strain, time and temperature would be extremely complex, especially since the values of any of the four parameters depends not only on the other three, but also on their history. For most practical purposes it is sufficient to look at the effects of temperature by comparing the results of creep tests carried out at different temperatures. Figure 8.8 shows the effect of temperature on stiffness, or tensile creep modulus.

Creep modulus, defined as:

$$\text{Creep modulus} = \frac{\text{stress (constant)}}{\text{strain (time dependent)}}$$

It depends, unlike Young's modulus, on time for linear viscoelastic materials, or time and stress and/or strain level for non-linear viscoelastic materials. Since moduli are factors which often occur in engineering formulae, it is useful to be able to produce creep modulus/time curves. These can be derived from the isometric stress/time curves by dividing the stress at any point by the strain.

We now need to introduce two further associated factors, both of which were discussed in Chapter 2. These are: strain recovery and stress relaxation.

8.5.4 Strain recovery

On removing the load at the end of a creep test, the recovery of deformation in a metal is generally confined to the recovery of elastic strain, and this occurs almost instantaneously. A viscoelastic material—provided it has not been stressed beyond its yield point—is able to recover the strain produced even after long periods of creep. Like creep, the recovery of strain in thermoplastics is time dependent. In general, recovery after short times at low strains tends to be faster than after long times at high strains. Figure 8.9 shows that recovery in acetal copolymer can exceed 90% when the specimen has recovered for a period ten times the initial creep period.

8.5.5 Stress relaxation

In many applications where the material is subjected to an applied stress at constant strain; for example, where an interference fit is involved, such as in pipe couplings, plugs, closures, etc., the stress in the material will gradually decay. This type of behaviour has already been discussed in Chapter 2 and is represented by the Maxwell Model (Fig. 2.11). As a first approximation we can obtain an idea of the extent of stress relaxation with time by sectioning through the basic family of creep curves (Fig. 8.5) at the appropriate strain value parallel to the time axis. If more precise results are required stress relaxation data should be obtained separately, since under certain conditions there are considerable differences between creep modulus and stress relaxation modulus.

8.5.6 Other long term stress systems

Strength, like deformation, is time dependent and, therefore, for design purposes single 'strength values' are of little real use; a series of tests to failure should be carried out and the results treated in a similar way to those already discussed for creep data.

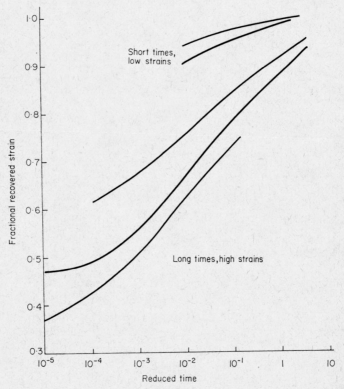

FIG. 8.9. Recovery from tensile creep of acetal copolymer at 20°C and 65% RH. (Courtesy I.C.I. Ltd.)

As we observed earlier the information most easily obtained is that measured under uniaxial tensile creep conditions. However, two other stress systems are of possible importance, although so far little useful data are available. These are compression and shear. The modulus in compression is generally found to be larger than that measured in tension and, as a useful approximation, the value of shear creep modulus is obtained by dividing tensile creep modulus figures by three.

Excessive approximation is dangerous however, like excessive extrapolation, and one must at all times be aware of the possible effect of such complicating factors as the presence of absorbed moisture, or a 'hostile' environment on the behaviour pattern of the material under stress.

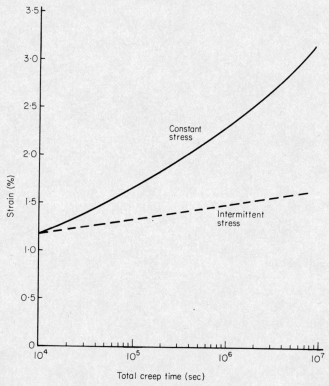

FIG. 8.10. Tensile creep of a polypropylene homopolymer under intermittent stress; 10 MN m^{-2} at 20°C for 6 hr each day.

8.6   SHORT TERM STRESSES AND FAILURE

So far we have discussed the behaviour of materials which have been subjected to a load which has remained constant throughout the service life of the component. However, there are occasions when the component is subjected to a load which varies with time. When very short times are involved the phenomenon is described as impact loading. Alternatively, if the load is applied intermittently the component suffers 'fatigue'. There is a continuous spectrum of response as a function of time or straining rate which will become apparent if we consider some typical behaviour patterns.

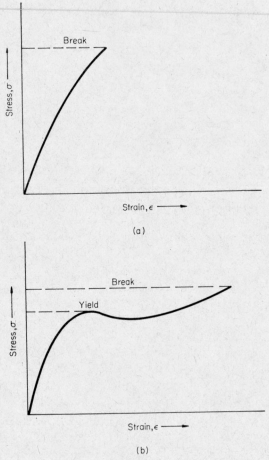

FIG. 8.11. Stress/strain curves at constant strain-rate. (a) 'Brittle' plastic. (b) 'Tough' plastic.

8.6.1 Periodic loading

If a component is subjected to relatively short periods of stress each followed by an unloaded period, we find that creep is less severe than that produced under continuous loading at the same stress. The recovery rate for viscoelastic materials decreases with increasing time; therefore doubling the unloaded period will not produce a comparable improvement in recovery. Nevertheless the type of performance we might expect from a propylene homopolymer is indicated in Fig. 8.10.

An important practical value of the results shown in Fig. 8.10 is that the amount of material required to perform a given function under periodic loading conditions is significantly less than that required to carry out the same function under constant load.

8.6.2   The strength of polymers

Some plastics materials under given conditions behave in a brittle manner and their uniaxial tensile stress/strain curves are of the type shown in Fig. 8.11(a). Here tensile strength is synonymous with stress at fracture, and this corresponds to the limiting strength of the material under similar conditions. For plastics which behave in a ductile manner, there is another important stress (cf. Chapter 2) which corresponds to the onset of large permanent deformations (Fig. 8.11(b)). Both of these parameters are strongly affected by changes in time and temperature and also by the rate at which the material is strained.

The effect of temperature is particularly noticeable under fatigue conditions since, unlike metals, plastics have a low thermal conductivity and exhibit high mechanical hysteresis. Thus, unless the applied energy per cycle is small, there will be a significant rise in temperature which will cause a decrease in resistance to deformation.

The effects of time, temperature and strain rate on typical materials are shown in Fig. 8.12.

8.6.3   Impact behaviour

Although the most satisfactory material for most applications is one that behaves in a ductile manner over the widest range of conditions, experience has shown that plastics which fail under impact conditions almost always do so in a brittle manner. The transition from tough to brittle failure is encouraged by several factors which include:

(a) decrease in service temperature;
(b) increase in straining rate;
(c) processing variables such as orientation;
(d) stress concentrations due to sharp changes in cross-section, corners, etc.;
(e) 'molecular' considerations including molecular weight, molecular weight distribution, branching and additives.

In order to test for these effects, it would be possible to induce brittle behaviour by increasing the severity of the tensile test through an increase in speed, reduction in temperature, or the inclusion of notches to provide stress concentrations. However, such a test is really too sophisticated and modifications of the pendulum testers used for metals have been successfully applied to plastics. Both the Izod and Charpy machines are used (the former in accordance with BS 2782:306A or ASTM D256) and some work has been carried out by a number of workers on a type of fatigue impact test in which the material is subjected to a series of impacts.

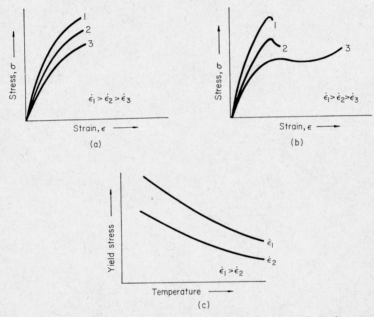

FIG. 8.12. The effect of a variety of parameters on the strength of plastics materials. (a) Stress/strain curves at different strain rates for 'brittle' plastic. (b) Stress/strain curves at different strain rates for 'tough' plastic. (c) Yield stress *vs* temperature at two different strain rates.

8.7 PROCESS CONSIDERATIONS

We have already discovered that, as is the case with metals, there is a considerable range of methods by which polymeric materials may be fashioned into the required product shape. All of these are likely to have their place, and their minimum economic level of production has been summarised in Table 8.1. Of particular importance to the design engineer are the production methods which often allow parts of complicated shape and intricate detail to be made in one rapid operation, with little or no finishing.

TABLE 8.6

Summary of the advantages and disadvantages associated with different manufacturing methods

Manufacturing method	Advantages and disadvantages													
	Capital machine cost	Tool/mould cost	Material cost	Output rates	Cycle times	Dimensional accuracy	Finishing stages	Thickness variation	Stress in mouldings	Can mould threads	Can mould holes	Open-ended components moulded	Inserts moulded-in	Waste material
Injection moulding	high	high	low	high	low or high	good	none	low	some	yes	yes	yes	yes	none
Blow-moulding	low	low	low	high	low or high	fair	some	fair	some	yes	no	yes[a]	no	some
Rotational moulding	low	low	high	low	high	fair	some	low	none	yes	no	yes[a]	yes	none
Extrusion	high	low	low	high	high	good	yes	high	some	no	no	yes	no	some
Thermoforming	low	low	high	high	high	poor	yes	high	some	no	no	yes	no	some

[a] Open-ended articles made by splitting closed moulding.

Most of the important processes have already been described in some detail in previous chapters, but as a rough guide to assist the designer some of the advantages and disadvantages of a selection of fabrication methods is presented in tabular form (Table 8.6).

It must also be remembered that many fabrication processes produce scrap, either as part of the primary process or as a product of the finishing or trimming operations. It is therefore necessary to be clear whether this

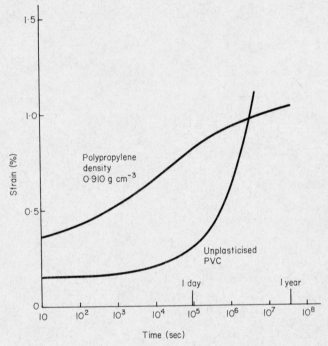

FIG. 8.13. Comparison of tensile creep curves at 60°C and 2 MN m⁻². (Courtesy I.C.I. Ltd.)

waste material—or a proportion of it—can be re-used, either in this or another process, or whether it has to be disposed of in some other way.

The factors which determine the selection of a particular method of fabrication must also include the various aspects which contribute to the manufacturing costs. These are:

(a) raw material cost;
(b) depreciation of machines;
(c) depreciation of tools;
(d) labour and other running costs and overheads.

8.8 USING THE DATA

Although at first sight the path to choosing the correct material for a particular application seems almost impossibly twisted, the procedure can be simplified and made quite manageable by following the relatively straightforward procedure outlined at the start of the chapter. Most important is to distinguish between what is essential and what can be tolerated, and to be aware of the particular problems and benefits associated with each of the fabrication processes which might be used. As far as the material is concerned, it is also necessary to give some thought as to whether the requirements can best be fulfilled by using the material singly or in combination with others, as discussed in Chapter 7. Once this has been provisionally decided, it is essential to study very carefully the projected long-term performance of the product and make use of the various creep curves and their derivatives. This is important because often the information gleaned by extrapolation is not merely vague but wrong. For example, Fig. 8.13 compares a PVC and a propylene at 60°C and 2 MN m^{-2}. At short times under load the PVC is better, but after a few months it suffers a sharp deterioration and the order is reversed. At 20°C the modulus of PVC is about 2·76 GN m^{-2} and that of polypropylene 1·17 GN m^{-2} and no indication of the cross-over is given.

In order to give an indication as to how the available information can be used most effectively for a specific application, it may be helpful to use as an example a design brief for an impeller. (The schedule of requirements has been kindly provided by I.C.I. Ltd.)

The design brief for a mechanical performance should cover at least the points shown in Table 8.7. The full design brief would, of course, also

TABLE 8.7

Design brief (mechanical)

---

1. Limitations on size, shape and weight.
2. Limitations on deflection, distortion or deformation.
3. Mechanical
   Steady loads:
   Magnitude and duration
   Fluctuating loads:
   Magnitude and frequency.
4. Temperature
   Steady temperature and time at that temperature
   Time at maximum temperature, and frequency
   Time at minimum temperature, and frequency.
5. Likelihood of accidental damage.
6. Chemical resistance requirements.
7. Exposure to sunlight, weather, etc.

---

include aesthetic, ergonomic, marketing and other functional requirements. To be more specific still, the design brief for the impeller should also include the technical points shown in Table 8.8.

TABLE 8.8

Design brief for an impeller

1. Dimensional accuracy needed (will it hit the casing?).
2. Maximum speed.
3. Maximum ambient temperature (in storage or tropical sunlight).
4. Maximum temperature at maximum speed.
5. Duration of speed and temperature (whether intermittent or continuous duty).
6. Average expected life under these conditions.
7. Corrosion conditions (solvents, acids or cleaning solutions).
8. Erosion conditions (sand and water).

FURTHER READING

Thermoplastics and Mechanical Engineering Design (9117), I.C.I. Plastics Division, Welwyn Garden City, Herts.
PINTEC 1972 (Conference Papers), Plastics Institute, London.
Shell Polyolefins Engineering Design Data, Shell Chemicals, 1966.
Design Engineering Series, *Plastics* 1 & 2, Morgan-Grampian, London.
Roff, W. J., and Scott, J. R. (1971). *Fibres, Films, Plastics and Rubbers*, Butterworths, London.
Ogorkiewicz, R. M. (Ed.) (1970). *Engineering Properties of Thermoplastics*, Wiley, New York.
Turner, S. (1973). *Mechanical Testing of Plastics*, Iliffe, London.

INDEX

Acetal resins, 20, 204
Acrylic
  fibres, 183
  polymers, 20, 196, 204
  rubbers, 62
Acrylonitrile/butadiene/styrene
    (ABS) polymers, 24, 25, 60,
    196
Addition polymerisation, 22, 23
Additives, 56, 57, 228, 229
  antioxidants, 62
  antistatic, 62
  asbestos, 65, 222, 226
  carbon black, 61, 63, 225, 227, 228
  carbon fibres, 225, 232, 235, 238,
    246, 247
  fibrous, 64, 65
  gaseous, 58
  glass fibres, 65, 222, 225, 230, 231,
    232, 233, 234, 235, 238, 239,
    240–246, 248
  liquid, 59
  particulate, 62, 64, 225, 226, 227,
    228, 233, 236, 237, 238
  polymeric, 60, 61, 62, 63, 225,
    227, 228, 230
  rubber, 225, 227, 228, 230
  stabilisers, 62
Air knife, 155
Alkyd polymers, 16, 17
Alloys, 4, 6, 9, 24, 25, 26
Amorphous polymers, 12, 14, 32, 48,
    49
Anelasticity, 41, 44, 45
Antioxidants, 62

Antistatic additives, 62
Asbestos, 9, 24, 65, 222, 226
Assembly methods, 207–212
Atom movements, 21, 36, 37, 38, 41,
    42
Atomic bonding, 31
Atoms, 3, 9, 10
Autothermal extrusion, 122

Banbury mixer, 57
  see also Mixers, internal
Benzene ring, 11, 12
Biaxial orientation, 13, 14, 15, 154
Blends, polymeric, 62
Blow moulding, 137, 163–169, 258
  methods of, 169, 170
  parison control, 140, 167, 168
Blowing agents, 58
Blown film (tubular), 149–154, 157
Bond strengths, 9
Brass, 4, 5, 6
Breaker plate, 136
Bulk modulus, 30, 31, 34, 234
Butadiene polymers, 60, 62, 228, 230

Calender, 29, 171–181
  faults, 180
  feed methods, 179
  operation, 179, 180
  roll arrangements, 171, 172, 173,
    175, 176, 177
  temperature control, 177, 180